Human Diversity

HUMAN DIVERSITY

Richard Lewontin

SCIENTIFIC AMERICAN LIBRARY

Scientific American Books
An imprint of W. H. Freeman and Company
New York   San Francisco

Library of Congress Cataloging in Publication Data

Lewontin, Richard C., 1929–
    Human diversity.

    Includes index.
    1. Human genetics.   I. Title.
QH431.L4186   1982   573.2′1   82-16723
ISBN   0-7167-1469-8
ISBN   0-7167-1470-1 (pbk.)

Printed in the United States of America

Scientific American Library is published by
Scientific American Books, an imprint of
W. H. Freeman and Company, San Francisco.

1 2 3 4 5 6 7 8 9 0 KP 0 8 9 8 7 6 5 4 3 2

To Mary Jane

". . . nor custom stale her infinite variety"

Contents

Preface

The extraordinary diversity of form and face that characterizes human beings has, over and over, been a source of wonder and delight. Pliny the Elder in his *Natural History* remarked that, although the human countenance was made up of only "ten parts or a little more," people were so fashioned that "among many thousands no two exist who cannot be distinguished." Nor is an individual person constant throughout life. From birth to death each one of us is in a perpetual state of flux in all our aspects. How are we to explain this rich multiplicity of human physical and psychic forms? The most dramatic and notorious attempts at explanation have been those simplistic theories that we know as genetic determinism and behaviorism. For the genetic determinist, all differences in form and function between people are to be referred, ultimately, to differences in their genes. For the behaviorist, psychic differences are simply the learned responses to repeated stimuli. But the truth is more complicated and less dramatic. An understanding of human physical and psychic variation requires a synthesis of molecular biology, genetics, development, physiology, psychology, sociology, anthropology, political economy, and history. So when, at the instigation of Gerard Piel, a group of us formed a plan to write a series of books that would illustrate basic scientific principles by their application to different problems, the subject of human diversity seemed an immediate choice.

Obviously, a book of fewer than 200 pages, written by a single, intellectually limited author cannot possibly bring together in proper balance and with sufficient depth of understanding even the very limited knowledge that exists on human psychic and physical development. It seemed to me, however, that we do know enough about genetics and development to discredit the naive determinism that has so often passed for science and to provide a hint of the real complexity behind human variation. At the moment, too little is known of the development of mind and body to make a definitive explanation, but there is no reason to suppose we will not eventually understand human diversity fully. As the apostle Paul wrote to the Corinthians, "For now we see through a glass darkly; but then face to face: now I know in part; but then I shall know even as I am known."

In describing what is to be seen through that glass, I have been greatly aided by the perceptive work of an anonymous editor, who insisted that I use words accurately and get facts straight, and by the equally perceptive work of Amy Malina, who assembled the illustrations. I am greatly indebted to Elizabeth Adkins Regan of Cornell University from whose work I borrowed heavily in my description of the origin of sex and gender differences. The entire work of editing, illustrating, and producing *Human Diversity* was overseen by Patricia Mittelstadt and by Linda Chaput, who also undertook to reassure me that my prose was not all *that* bad. On the other hand, Becky Jones, who with her accustomed care and intelligence typed and assembled the manuscript, convinced me that my handwriting *was* all that bad. The most important part of the book, the index, was made with her usual perspicacity by Mary Jane Lewontin.

RICHARD LEWONTIN
Marlboro, Vermont
July, 1982

Human Diversity

Human Variety

I

When we think about human beings, we are struck with an apparent contradiction in the nature of our own species: Human beings are alike, yet they are all different. When, disgusted with some friend's petty behavior, I say "Ah, well, Mary Jane, I guess people are all alike," I am contradicting myself. People cannot literally be "all alike" or I would not have been able to pick out Mary Jane from everyone else.

There are indeed remarkable similarities among all human beings. These similarities constitute a "human nature" that binds us together and differentiates us from all other organisms. Our erect posture, thin skin, and relative lack of hair mark us off physically from all other mammals. All organisms transform, to some extent, the environments in which they live, but we are unique in our radical alteration of the world to fit our own needs. Moreover, the means we use to change and control the world, our rich and intricate social organization, and our use of abstract language are purely human attributes—they set us off clearly from the rest of organic nature. It is fashionable to talk of "insect societies" and "chimpanzee language," but the concepts *society* and *language* have been derived from human experience and have been applied only secondarily to phenomena in the lives of other organisms. Indeed, the simplicity of the interactions in "insect societies," as compared with our own cultures, and the poverty of chimpanzee "language" reinforce our sense of human uniqueness and superiority. After all, it is human beings who write books about insect societies and who teach chimpanzees to press buttons, not the other way around. It is precisely in the contrast with other species that we come to perceive the common features of humanity. The differences among human beings virtually disappear from view in the immensity of the contrast between us and other animals, even our nearest primate relatives.

Despite the similarities that appear when we observe human beings from a philosophical distance, our everyday experience reveals that there is an extraordinary amount of variation among us, too. We differ in height, weight, hair texture, skin color, facial features and expressions, posture, gait, and costume. We have not the slightest difficulty in picking out the face of a friend from a large crowd of faces, even at a glance. We know hundreds, if not thousands, of names that we can associate with particular people on sight. Even identical twins who act and dress alike can be readily told apart by their parents, their siblings, and others who know them well. But we do not need the wealth of information provided by complex visual images in order to differentiate one person from another. Most of us can recognize our friends and some of our acquaintances by their voices alone.

Variation as a Social Product

It is obvious that our ability to perceive individual variation is partly determined by our social conditioning. After all, perception is a subjective phenomenon. Perhaps less obvious is the fact that *objective* variation among people is also socially conditioned. Western society, especially since the English, French, and American revolutions of the seventeenth and eighteenth centuries, has placed a great emphasis on individuality and the importance of the unique personality of each human being. Western society places great value on the freedom of people to sell their labor power in a competitive market and to rise or fall in the social hierarchy according to properties of intellect, drive, will, and skill that are supposed to reside in each person.

This was not always so, nor is it universally true today. European feudal society was much more collective and more organic in its structure than modern capitalistic society. For the most part, one's place in feudal society was preordained and stable. That place was determined not by one's personal qualities but by customary relations among people playing customary roles—peasants, artisans, clerics, and landowners. There was very little social mobility; throughout their lives, people played stereotyped social roles determined by their social class. The emphasis on individuality that seems such an unquestioned part of our social reality was absent. Even today, many societies (the Navajo, for example) disapprove of individualistic or idiosyncratic behavior. For them, a nonhierarchical collective social organization is the mode. In contrast, modern Western society is characterized by the ideology of individuality. In novels and poetry, newspapers and political speeches, in the schools and in sermons, the individual is exalted. This ideology makes us more sensitive to variations among people; it also actually creates variation by putting a positive social value on

being different—within certain acceptable limits. For example, people vary their dress, hair styles, furnishings, cars, houses, facial expressions, and voices to give themselves a sense of being individual, of being "just a little bit different."

Despite the immense social weight placed on individuality, we are acutely conscious of the similarities within groups. We recognize men and women as distinct groups, and, despite the myth of the classless society, we have no trouble distinguishing miners from mine owners. In addition to sexual and social distinctions, there are clear geographical differences. The Finns resemble each other more than they do the Italians. Generally, they have lighter skin and hair, and, of course, they speak a very different language. But Finns and Italians resemble each other more in shape, color, and culture than either group is like the Australian aborigines. Human geographic variation is thus organized in a kind of spatial hierarchy. Members of a local ethnic group or tribe resemble each other more than they do members of other groups living in the same general geographic area. There are, in turn, differences between the peoples of different major geographical areas. Thus, an understanding of human diversity depends upon an understanding of how such differences between groups arise and how large they are relative to the differences between individual members of a group.

Sex Differences

The most obvious group difference, the one to which we are exposed every day and of which we are intensely aware, is the difference between males and females. Obviously, a major difference between the sexes consists in dissimilarities in the structure of their reproductive organs. Although these characteristics, which are established in the embryo, are the ultimate basis of sex identification, it is not these primary sex differences that we ordinarily use to distinguish men from women. Instead, we depend in part on secondary sex characteristics—such as degree of breast development, coarseness of facial hair, and differences in skin texture—which begin to develop in adolescence and which we expect to see as a normal part of sexual differentiation.

In part, we also depend on what are better called *gender* differences, rather than *sex* differences. These are differences in attitude, interest, social dominance, and knowledge that develop in children during the course of their socialization and that bear an uncertain relation to primary and secondary sex characteristics. Sometimes a major shift in a gender difference will upset our expectations and confuse us about the sexual identity of the people among whom it has occurred. For example, in the 1960s, when young men began to wear their hair very long again (the practice had been out of fashion for several generations), many people were confused and disturbed by what seemed to be conflicting signals about sexual identity.

The everyday practice of making generalizations about differences between the sexes—saying, "Men are taller than women," or, "Women are more sensitive than men"—obscures the fact that there is immense variation within sexes in all physiological, psychosocial, and anatomical traits. There is, moreover, a great deal of overlap in the ranges of these traits. Many men are shorter and lighter in weight than many women; many women have deeper voices than many men; and many women are more dominant and aggressive than many men. Some men have larger breasts than some women, and numerous men have no coarse facial hair at all. Yet, despite the widespread overlap of characteris-

The "normal" length of men's hair changes from time to time and differs between social classes. How shocked we were at the Beatles, yet how accepting of Einstein's longer locks.

tics, we usually can clearly identify a person as male or female. To be sure, some cases are ambiguous, and it is a sign of the deep importance we place on sexual assignment that we find such people disturbing and puzzling.

As is true of individual differences, modal differences between the sexes are, in part, our inventions, based on some previous expectation. In the famous "Baby X" experiments carried out at City University of New York, three sets of adults were asked to play with and describe babies of different sex, dressed in yellow. When the adults were told the babies were girls, they offered the infants dolls to play with and described the babies as "fragile" and "gentle"; if the children were said to be boys, they were given footballs to play with and described as masculine. When no information was given about their sex, the adults tried to find out. On being given an evasive answer, they assigned a sex to each child and justified their judgments by describing the "boys" as having a strong grip and the "girls" as being soft, fragile, and cuddly. In fact, the adults did no better than random in their guesses. Clearly, prior sexual identification creates seemingly objective differences in the eye of the beholder. These differences can, in turn, affect the actual development of the child who will conform to some degree to those expectations. It is a characteristic of human psychic development that the subjective becomes the objective.

Sex and gender differences are especially susceptible to reinforcement by social conventions. Men and women dress differently and wear their hair in different styles, although there is no objective necessity for doing so. We move our bodies in "masculine" and "feminine" ways. Voice pitch, which is partly a consequence of morphological changes that occur at puberty, is modulated—at first consciously, and then unconsciously—to conform to masculine or to feminine stereotypes. Nearly all of us have been put under pressure, at one time or another, to "act like a man" or "be more ladylike." What begins as a difference in plumbing fixtures becomes, through an intricate system of social reinforcement, a complete remodeling of the building.

The Social Division of Labor

In some societies, the division between the sexes closely corresponds to a division of the various forms of productive labor. Among the Yoruba of western Africa, men do the bulk of farming tasks; among the Dani of New Guinea, however, women are the farmers. Tuareg men of northern Africa tend the camels, but women milk them. However, in technologically advanced societies, such as those engaged in large-scale sedentary agriculture, the division of productive labor cuts across sex lines and has given rise to social classes and to occupational groups within those classes.

A great deal of actual human diversity is a consequence of the social organization of productive activity. The factory system introduced in Europe at the end of the eighteenth century resulted in an increasing division of skill, knowledge, and labor among different people. Whereas the artisan weaver was capable of performing all of the operations of cloth production from yarn to undyed cloth, a modern textile mill separates bobbin fillers from warpers, from weavers, from machine repairers, and so on. The process of subdivision of tasks still continues, as employers strive to reduce the cost of labor and to increase their control over the labor process itself. An important consequence for the workers is that they lose control, understanding, and a diversity of skills. When computers were first introduced, a single person might know how to punch cards, how to write a program, how to wire control panels, and how to run the computer. Now machine operators, data clerks, systems programmers, and applications programmers are all separate people who differ in knowledge, skill, education, salary, prestige in the workplace, and lifestyle. Moreover, the differentiation of labor reinforces the differences between the people who do the different jobs. The demand for a highly differentiated work force results in a more highly compartmentalized educational scheme with more "tracking" and more extremely specialized task training. As a result, people become, objectively, more different from one another than they once were: What is manifest as a variation in knowledge, skill, understanding, and lifestyle among people is a consequence of varied developmental experiences that have, as their purpose, the production of people with different knowledge, skill, and understanding.

My father-in-law never went to high school. He learned about wireless telegraphy in the U.S. Navy and took correspondence and night courses in simple mathematics and physics. As a civilian, he helped to design and construct a major overseas telephone transmitting and receiving station, and then he ran it for many years. He knew the elementary theory behind the workings of his equipment, and he could diagnose breakdowns and repair them. He became an expert on the occurrence of magnetic storms caused by sunspots. He was the treasurer of his church. He was what was known in the nineteenth century as "an aristocrat of labor." His radiotelephone station has long been shut down, and its rudimentary equipment has been discarded. Such installations are now completely designed by electronic engineers with graduate degrees. Their operating schedules are determined by computers tended by people who know only a few routine mechanical tasks. Repairs are made by such workers, who plug in new integrated circuits under the supervision of an engineer. There are those who know and those who do. The self-educated, skilled aristocrat of labor has been replaced by a highly trained professional engineer and a set of minimally trained manual workers.

An interesting question, one to which this book is in part addressed, is to what extent such developed variation between people may reflect (and be built upon) innate biological differences between them. Answering that biological question, however, will not tell us why our society shows such an extreme differentiation of skill and labor and status. We may suspect that a virtuoso violinist is born with different neuronal connections than those of us who cannot carry a tune, but the soloist playing for an audience of hundreds, for a fee of thousands, whose name is known to millions is a socially constructed creature. In the seventeenth century, a person of such skill took his meals with the footmen and grooms, who were his social equals.

Geographical Variation

Anyone who has not lived his or her entire life in an isolated valley, cut off from all communication with the rest of the world, is aware that people from different geographic regions are different. It is an unquestioned part of our perception of the world that there are black, white, brown, or yellow people, and a lot of other people who fit somewhere in between. Moreover, we make finer distinctions. Among those people we perceive to be white, we recognize many differences, including those between blond Finns and much darker Sicilians.

As a consequence of the large-scale movement of people during human history, groups that were originally separated geographically now live side by side. Europeans invaded the original homelands of New World Indians. Black slaves from Africa followed soon after. In the Wooloomooloo district of Sydney, Australian aborigines, the English and Irish who took Australia from them, and the more recent Italian, Greek, and Maltese immigrants live side by side, yet separately. There are many objective differences between groups from different regions. No one would confuse a Pygmy from Zaire with a Watusi from East Africa. Yet, even at this level of human differentiation, social relations modulate our subjective impressions of human differences and the objective reality itself.

Although we have not the slightest difficulty in differentiating among individual persons within our own group, "*they* all look alike." The differences in skin color, facial form, hair texture, posture, and costume between Europeans, Asians, Africans, native Americans, Oceanians, and other large groups seem so

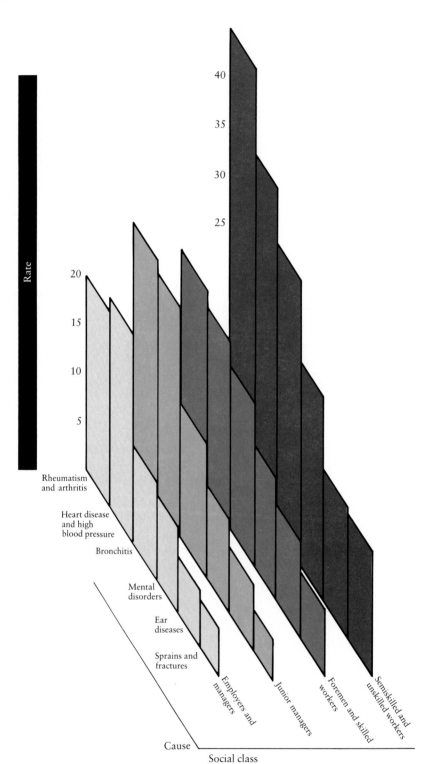

Class differences in the cause of disease in England. (Rate is the number of cases per 10,000 persons.)

broad that the individual variation within these groups may be lost to an outside observer. Once, in upper Egypt, my wife was approached in our hotel lobby by a stranger, an Egyptian, who began to discuss some completely mysterious matter with her. She insisted that she did not know him, and he, finally, upon looking around, spied another person he had mistaken her for: another foreign woman who, needless to say, looked nothing like my wife. He apologized for the intrusion, saying, in effect, "I am truly sorry, but you all look alike."

Differentiating among the members of an unfamiliar group is not simply a matter of opportunity and practice; it is a question of attention and intention. White Americans, although they may see thousands of black Americans each day in cities where both work, continue to have difficulty in telling black people apart. Unless we wish to assume that whites are biologically deficient in their ability to perceive differences that are obvious to any black person, we must conclude that these whites simply do not confront blacks as individual persons because of the social significance of race in the United States. Although blacks and whites share the sidewalks and subways in New York City, they live in different "social spaces."

A consequence of the differentiation of social spaces is that objective differences between groups develop and are perpetuated. No one doubts that there are real differences—in language, tastes, skill, knowledge, wealth, and power—between groups and social classes throughout the world. Some of these differences developed historically as a consequence of the isolation of one group from another—as did, for example, the major linguistic differences. Yet others are precisely the result of the relations between groups—for example, the differences in social power between groups within a single society. What is less obvious is that seemingly purely physical differences may arise in the course of social interactions. Variations in posture and gait, in facial expression, and in pitch and intensity of voice are all evident between racial and ethnic groups and social classes. Mimes (and other actors) learn how, with a few tricks of gait and posture, to convey the differences between people of the lower and the upper classes or between stereotyped whites and blacks.

Differences in patterns of sickness and mortality, too, are characteristic of socially defined groups. The probability of falling sick from a variety of causes depends on one's occupational status. The graph on the facing page shows, for Britain, the differences among managers, professionals, and blue-collar workers in patterns of morbidity. Chief executive officers of textile mills are not known to contract brown lung disease, although the disease is relatively common among the workers in those mills. There is virtually no aspect of the variation among human beings that is not in some way influenced by the social organization that characterizes our species.

Human Variation

Our consciousness of the variety of human beings compounds variations that are biological with those that are cultural, and it is not always easy to distinguish the roles of the two. A Swedish child differs from a Japanese child not only in hair color, eye color, and skin color, but in facial expression, posture, and, of course, costume. The shape of a mouth, the openness of eyes, the smile and frown lines that appear as biological features of shape are in fact superimposed on our faces by our unconscious mimicry of those around us. Europeans find it extremely uncomfortable to sit on their haunches, yet until the spread of European culture throughout the world they were the only people to sit "hanging their legs," as the Japanese called it. A style of beard, a haircut, a nose ornament, a face covering—all accentuate what seem to us to be "natural" variations that we associate with different ethnic groups.

Genes, Environment, and Organism

2

The rich diversity of human beings, both within and between groups, has its roots, ultimately, in human biology. At the simplest level, that statement is true because people differ from one another biologically: We are born with different sets of genes. At a deeper level, our biology is the wellspring of our diversity because it is human biology that has created both the possibility of human society and the necessity for it. Human society is possible only for an organism with a nervous system as extraordinarily well developed as the human nervous system. (But it is also possible only for organisms of a reasonable size. Swift's Lilliputians are necessarily a fiction because six-inch manikins could not lift a heavy enough instrument high enough to generate the kinetic energy needed to break rocks and extract minerals. Nor could they control fire because the tiny twigs they might carry as torches would be consumed in an instant by the flames.) Human biology makes human society necessary because we are helpless for a very long period during infancy and early childhood. Children of, say, five years could never have survived in isolation in the past any more than they can today. Some social organization of food gathering and production, of protection of the vulnerable, and of learning about the world is essential for the survival of organisms as physically fragile as we are when we are children.

The problem of understanding the sources of human diversity is, in essence, the problem of relating the biological variation that individual people inherit from their biologically diverse parents to the effects of the environment in which those people live. This problem is often stated in terms of a contrast between "nature" and "nurture." Do I differ from my neighbor because I have different genes or because my life experiences have been different? Is the difference between our scores on an IQ test a result of the differences between our genes or of the differences between our environments? At first sight, these seem to be sensible questions. Common sense tells us that the difference in skin color between a European and a sub-Saharan African must be innate: The children born to European colonial families in Africa were still white, and the children born to slave families in North America were still black. At the other extreme, it is also common sense that the ability to speak English—as opposed to, say, Polish—is totally socially determined. These observations may lead us to ask what seems to be a reasonable question: "Is the variation among people, with respect to a given trait, the result of nature or of nurture?" It turns out, as we shall see, that such a question poses the problem of the causation of differences in a completely invalid way. Despite its common-sense appearance, that question is, in fact, biological nonsense. Sometimes the problem is posed in what appears to be a more quantitative form: Because neither nature nor nurture can be solely responsible for determining a trait, we might ask, "What are the relative roles of heredity and environment in determining a trait?" Thus, it might be said that IQ

differences are 80% genetic and 20% environmental. Such a pluralistic approach to causation is as incorrect as the first question, and it arises from the same misunderstanding about biology: The error is in the attempt to assign separate causal roles to internal and external forces in the formation of individuals and society.

The attempt to make a clear separation of internal and external causes goes back to the machine model of organisms, created by René Descartes in the seventeenth century. For Descartes, living beings, like the physical world, were clockwork mechanisms that could be understood by taking them to pieces and studying the individual parts. This Cartesian method of analysis has been the basis of virtually all progress in biological science. Physiology, genetics, and molecular biology proceed by breaking things down into finer and finer parts and then attempting to reconstruct the whole functioning organism by putting them back together again. The method, in turn, reinforces the notion that complex effects are simply the consequences of the interaction of discrete, isolable causes. Thus, our bodies are seen as bundles of organs, each performing some particular function, and when they malfunction we are attended by specialists. Is your heart not working well? See a cardiologist. Is it your bowel? Try a gastroenterologist. Your brain? A neurologist, then.

The Cartesian approach has been very successful, but it has had its conspicuous failures as well. The problem of the functioning of the brain has been disturbingly resistant to solution, despite years of studying single neurons in isolation. Too, we seem to be no closer to an understanding of how a fertilized egg becomes a thinking, acting person than we were at midcentury, despite extraordinary progress in molecular biology. Partly as a consequence of such failures, biologists have become conscious of two fundamental features of living organisms. First, organisms are "open systems," constantly incorporating into themselves new material and energy from outside, constantly changing, constantly developing. Although I appear to have the same nose today that I had last year, in fact the molecules that make it up have been discarded and replaced in the interim. Second, internal and external factors cannot, in general, be assigned separate roles in determining the organism. A proper understanding of the origins of human "nature" and human diversity, then, rests on an understanding of these two fundamental features of organisms: first, that each organism is the subject of *continuous development throughout its life;* second, that the developing organism is at all times *under the influence of mutually interacting genes and environment.*

No person can be characterized by a shape, a physiology, a behavior. Rather, each person is a history of shapes, physiologies, and behaviors, a history that begins at the moment of conception and ends only after death, with the dissolution of the body into its constituent elements. Thus, the proper description of any given person is not a single characterization, nor even a set of characterizations, but a set of characterizations in a particular temporal order, a developmental history. There is no characteristic of any human being that is not in a continuous state of developmental transformation throughout life. For example, even our bones and cartilage are living materials that are constantly being recycled, disappearing and being replaced. As a result, a real shrinkage in height occurs in old age, quite apart from any changes in posture, although we are accustomed to thinking of height as being fixed at the end of adolescence.

Some changes in physiology are immediate, reversible responses to changes in environment. We perspire when hot, and our hearts pump faster when we exert ourselves, but these adaptations to stress disappear within minutes after the environment changes again. Some adaptive changes, however, reverse more slowly, and some are irreversible. Peruvian miners working at high altitudes develop enlarged hearts and enlarged chest cavities, and their bodies produce more red blood cells than they do at low elevations. The increase in blood-cell production takes place over a period of weeks and is reversed over a similar period when the miners live at low altitudes. The enlargement of their hearts and chests takes place over a period of years and is irreversible.

Two people who may be identical in some trait at one moment in their lives may later become irreversibly different. All Caucasian babies are born with eyes of a light, undifferentiated color, but very soon they develop different amounts of dark pigment. Conversely, the darkening and eventual greying of hair with age may eliminate differences between people that were pronounced when they were younger.

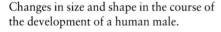

Changes in size and shape in the course of the development of a human male.

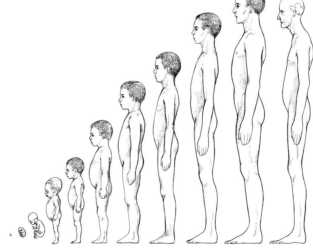

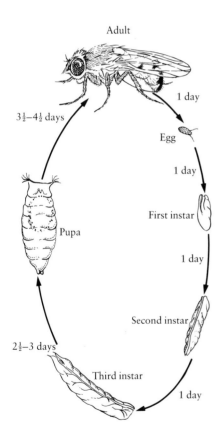

Adult

$3\frac{1}{2}$–$4\frac{1}{2}$ days

1 day

Egg

1 day

First instar

1 day

Second instar

$2\frac{1}{2}$–3 days

Third instar

1 day

Pupa

A fruitfly, *Drosophila*, with an extra pair of wings (top) and the life cycle of the fruitfly (bottom).

The nature of the developmental process is such that, at each instant of time, the change that is occurring is a function both of the present state of the organism and of the environment in which it finds itself. That is, to predict what an organism will be like at some future moment, it is not sufficient to know what it is like now, nor is it enough to describe the environment through which the organism is about to pass. We must know both. In this sense, the developing organism is like a space capsule. To predict its position at some time in the future, we must know both its present position and the forces of acceleration that will act on it in the interim. An important consequence of this constant codetermination of the characteristics of the organism is that the same environment encountered at different stages of the life cycle will have different consequences. For example, exposing a developing fruit fly *(Drosophila melanogaster)* to ether fumes will cause it to develop a rudimentary extra pair of wings—provided that the exposure occurs about three hours after the fertilization of the egg. Exposure at a later stage will not work, even though the fly's wings do not actually develop until much later—after the egg has hatched into a larva, the larva has gone through three successive developmental stages (instars), and the organism has completely reorganized itself as a pupa (see the drawing at the left). On the other hand, the stage at which the pattern of veins on the wing is sensitive to alteration is during the course of wing development itself, when a brief high temperature shock (but not an exposure to ether) will interrupt the development of a specific vein.

Another consequence of the historical nature of organisms is that the way an organism reacts to a particular environment at a particular stage in its life history depends upon the environments of the past, which are somehow recorded in the organism's physiology and anatomy. If adult flies are exposed to a high temperature (say, 32°C) when they are still larvae, they will be able to withstand high temperatures as adults. In human beings, much information about past environments is stored in the immune system. Sometimes, this can have disastrous consequences. If a child is badly stung by bees, he or she may develop an extreme sensitivity to future bee stings so that the next sting causes the body to produce massive amounts of antibody, resulting in shock or even death.

The historical determination of an organism's present state and future prospects should make us think again about the meaning of such descriptions as "ability." To say that I have more mathematical ability than a certain friend who happens to be of the same age is to say literally that, at the age of 52 years, I know how to solve equations while he does not. But what would it mean to say that I was *born* with greater mathematical ability than he? Surely not that, at the age of 52 hours, I could solve equations while he could not. It might mean that, irrespective of our life histories, I would one day have greater mathematical ability. Yet that cannot be true either: If I had spent my life as a landless peasant in the highlands of Guatemala, my knowledge of mathematics would be minimal. On the other hand, it might mean that, if we both had had identical environmental histories that included adequate opportunity to learn mathematics, I would be the better solver of equations at the age of 52 years. Such a statement might or might not be true, but it is a statement about possible developmental futures for two newborn organisms, not a statement about their abilities at birth. Abilities, like all other aspects of the morphology, physiology, and behavior of an organism, are manifest traits that develop over the life of the organism. Their development is contingent not only on the environments to which the organism is exposed, but also on the historical sequence of those environments.

The fact that organisms develop is sometimes expressed in the metaphor of the "developmental program," in which biological development is analogized to a computer program with built-in instructions. Fertilization of the egg establishes the full set of instructions and simultaneously presses the button for execution of the sequence of commands. Metaphors of this sort must be applied with great caution, because they often carry implications that can lead us astray. If development is to be regarded as a sequential computer program, it is a program of a complex kind: First, it is a program that reads in new data constantly. Second, depending upon the data, the program may branch to alternative sets of instructions. In the language of FORTRAN programs, each step reads "IF . . . THEN. . . ." Third, the data themselves may alter the instructions or become a new set of instructions not previously included in the program. That is the significance of learning. The danger of the "developmental program" metaphor is that we may be led to think of the simplest sort of computer program, in which the final result is encapsulated completely in the original instructions and awaits only a mechanical unfolding. Thus we become, in the mistaken phrase of Richard Dawkins, author of *The Selfish Gene,* "lumbering robots" who are ruled by our genes "body and mind." Nothing could be further from the truth.

The second fundamental feature of living organisms is that their developmental histories are a consequence of the unique interaction between genes and environment, between the initial "program" and the constant input of data (including new instructions). For the geneticist, the basic distinction is between *genotype* and *phenotype.* The genotype comprises what is inherited, through the sperm and egg, at the moment of conception: a set of DNA molecules, the genes, which are contained in the nucleus of the fertilized egg. The phenotype, on the other hand, is made up of all aspects of the organism—including its morphology, physiology, and behavior—at some particular moment in its life. We do not inherit our phenotypes. They develop throughout our lifetimes partly as a consequence of our genotypes—but only partly.

There are several errors commonly made in describing the relation between the genes and the organism of which we must beware in looking at human diversity. They are (1) that genes determine the phenotype, (2) that genes determine capacity, and (3) that genes determine tendencies.

Genes Determine the Phenotype This most elementary error asserts that, if the genotype of the organism is given, its phenotype is fixed. In general, that assertion is false. As we have seen, nutrition, workplace and nature of work, and social experience all make for differences in phenotype. Identical twins, who are genetically identical, will develop quite different body shapes and metabolic rates if one lives at sea level doing light work while the other does heavy labor at an altitude of 10,000 feet in the mountains. Although the phenotype generally depends upon the interaction of genes and environment, there are, of course, some traits that have a simple one-to-one correspondence with genes. The human blood groups provide examples of such traits (see Chapter 3). A person with a genotype specifying blood type A will always have type A blood, irrespective of environment. But such completely genetically determined characters are the exception rather than the rule.

Genes Determine Capacity This error might be called the "empty bucket" metaphor. In this view, each of us is born a collection of empty buckets to be filled by environmental experience. Our genes determine the sizes of the buckets, which represent various traits. In a favorable environment, the buckets will be filled to overflowing, so that observed differences between individuals will reflect the innate differences in the sizes of their buckets. In a poor environment, however, all buckets will be nearly empty, so genetic differences will not be apparent. This metaphor has been used, for example, to analyze the causes of differences in performance on IQ tests. It is asserted that children in poor environments will perform poorly on tests but that, in enriched environments, their innate differences in capacity will show up. In fact, however, there is no evidence that different genotypes have different capacities, in the sense of determining different maximal performances in "enriched environments."

It must be literally true, of course, that, for a given genotype over a fixed range of environments, there is some largest size, say, to which the organism can grow. But the environment in which an organism of one genotype grows to its largest size will, in general, be different from the environment most favorable for the growth of an organism of another genotype. For example, if plants from a population at sea level are grown at high altitude, they may barely survive, while plants from a high altitude will be stunted when grown at sea level.

All talk about the theoretical maximum phenotype for a given genotype really misses the point. The question is not whether one genotype could, under some conceivable set of circumstances, outgrow another. Rather, our interest is in the actual kinds of phenotypes that are developed by different genotypes in the various environments that exist or that can be created.

Genes Determine Tendencies This is the subtlest of the errors because it is framed in tentative terms and seems to take into account the environment in which an organism develops. To say that "Max has a genetic tendency to be fat" implies that, on some diets, but not on all, Max will grow fat. But, if there are some diets on which Max will be thin, why can we not describe Max as having a genetic tendency to be thin? The idea of tendency ideally carries with it the sense of "normally," or "usually," or "unless disturbed by outside forces." Newton said that bodies tend to stay at rest or in uniform motion "unless compelled to change that state by forces impressed thereon." To speak of genetic tendencies, we must have some notion of a set of environments that is usual or normal, so that any other environment to which the organism might be exposed can be regarded as a "force impressed thereon." When we look over the range of human environments, however, it is by no means clear how to distinguish normal from unusual environments as they affect important aspects of human variation. Is there any basis for saying whether high nutrition or low nutrition, the presence or absence of parasites, hard work or comparative leisure, individual striving and entrepreneurship or collective sharing are usual or unusual? Yet these variables have a profound effect on phenotype. People who "tend to be fat" on 5500 calories a day "tend to be thin" on 2000. Families with both "tendencies" will be found living in the same towns in northeastern Brazil, where two-thirds of the families live on less than what is considered a minimum subsistence diet by the World Health Organization.

We might interpret the notion of tendency as a comparison between a specific person and the average of the population. Thus, "Max has a tendency to be fat" might simply mean that, on 4000 calories a day, Max is fatter than the average person on a similar diet. But such a characterization would be useful only if Max were also fatter than the average when he and everyone else takes in 1500 or 2500 calories a day. Is Max fatter than the average no matter how many calories he and everyone else are taking in? The environment that increases the size of an organism of one genotype may increase the size of another much less, so that the order of the sizes of the two genotypes is reversed. Only if environments affect organisms of different genotypes in the same way is "genetic tendency" of any use as a way of describing genetic differences.

The Norm of Reaction

The basic concept for a correct understanding of gene and organism is the *norm of reaction*. For a given genotype, there will be a particular phenotype for each environment. The *norm of reaction* for a genotype is a list or graph of the correspondence between the different possible environments and the phenotypes that would result. Each genotype has its own characteristic norm of reaction. So, for example, a table giving Max's steady body weight for each possible caloric intake would be Max's norm of reaction for body weight in relation to calories consumed. For the same schedule of caloric intakes, Lillian's body weight would be different, showing a different norm of reaction. Rather than characterizing a genotype by a single phenotype or a single "tendency," the norm of reaction describes the actual relation between the environment and the phenotype for the genotype in question.

The eye size of three different genotypes of *Drosophila* (left below) is a function of developmental temperature. The numbers of cells (ommatidia) are given on a logarithmic scale. The scanning electronic micrographs are of the eye of a wild-type (middle) and a bar-eye *Drosophila* (right).

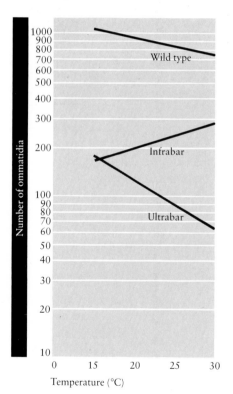

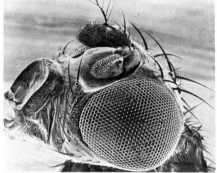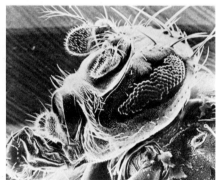

An example of such norms is given in the graph at the left. In fruit flies of the genus *Drosophila*, the eye is made up of a large number of identical units called ommatidia, each of which is a separate image producer (see the photographs above). The number of ommatidia in the adult eye depends upon the temperature at which the fly developed. Once the adult eye has developed, no further change in number can occur. The graph shows the relation between the number of ommatidia and temperature for flies of one genotype ("wild type") measured in an experiment. The relation is a slow logarithmic decrease in the number of ommatidia with increasing temperature. The graph also shows the norm of reaction for flies of a genotype containing the mutant gene *Ultrabar*. Again, there is a decreasing relation, but the flies with the *Ultrabar* genotype have fewer ommatidia than the wild-type flies at all temperatures and are more sensitive to temperature. Because flies with the *Ultrabar* genotype have fewer ommatidia than wild-type at all temperatures, we would not be distorting the truth to say that the *Ultrabar* genotype reduces the number of ommatidia. There is a third genotype shown, however: one containing the mutant gene *Infrabar*, which has a norm of reaction that runs in the opposite direction from that of the *Ultrabar* genotype and crosses it at low temperatures. Here we cannot say which genotype produces fewer ommatidia—it depends upon the temperature. At some temperatures, the number for *Ultrabar* is larger than that for *Infrabar;* at others, it is smaller. At one temperature, the numbers are identical. The genotypes cannot be compared by saying they "determine" different eye sizes, nor that one has a greater "capacity" than the other, nor that one "tends" to produce larger eyes than the other. The only correct way to describe the differences among the genotypes is to give their norms of reaction.

These norms of reaction included those of two laboratory mutants chosen to illustrate the possible relations between norms of reaction of different genotypes. It is very unusual, among genotypes that occur in natural populations of organisms, to find one genotype whose norm of reaction is constantly above others over a range of environments. The usual situation is illustrated in the series of drawings on the facing page, which consists of tracings made from actual photographs of plants grown in an experiment on reaction norms. Seven individual plants of a California herb of the genus *Achillea,* each of a different genotype, were collected from a population in the wild, and each plant was cut into three pieces. One piece of each plant was planted near sea level (lower drawing), one at 1400 meters (middle drawing), and one in the high mountains at 3050 meters (upper drawing). The three plants that grew from the pieces of each of the original plants are shown one under another, giving a pictorial norm of reaction for each of the seven genotypes over the three environments. As the drawings show, the tallest plant at sea level was not the tallest at the other two altitudes, and it did not even flower at the middle elevation. The order of sizes for the seven genotypes changes from altitude to altitude. None is consistantly the tallest or the shortest. Conversely, we cannot pick one of the environments as unambiguously the most favorable for all seven genotypes. No notions of determination, or tendency, or capacity have any meaning for describing the relations between genotype and phenotype. The phenotype is the unique consequence of a particular genotype developing in a particular environment.

The reader might remark that norms of reaction have been illustrated for flies and plants, but not for humans. Except for such traits as the presence or absence of blood-group antigens, which are absolutely constant across environments, we do not have the norm of reaction for any human trait. There are two reasons for this. First, in order to establish a norm of reaction, we must be able to expose organisms of the same genotype to a variety of environments during the course of development. That means that we must be able to specify and control the environment. It is easy enough to grow plants side by side in an experimental garden or to control the temperature at which fruit flies develop in the labora-

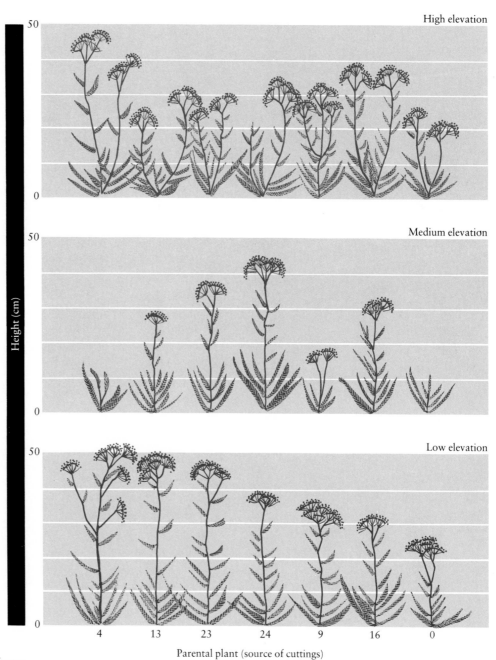

Growth of cuttings from seven different individual plants of *Achillea* when planted at three different elevations.

Monozygotic twins raised apart.

tory. But we do not even know which environmental factors are relevant to the development of most human traits, much less how to control them for an experiment. Certainly, early nutrition has some important effect on the height to which people grow, but it is remarkably difficult to get reliable information on people's actual food intake. For behavioral traits, the things that make up "personality," we hardly know what in the social environment we should begin to measure.

The second difficulty is more fundamental. In order to establish the norm of reaction for a genotype, we need to have many individual organisms of identical genotype, since exposing the same organism to different environments in sequence will not tell us what we want to know. We would need sets of identical quintuplets (or decuplets, or some such) whom we could separate at birth (or even before) and raise in controlled environments. In fruit flies, special controlled breeding methods are available that allow us to produce large numbers of genetically identical organisms. In *Achillea,* identical triplets were produced by the simple expedient of cutting each plant into three pieces. In humans, producing large numbers of genetically identical individual organisms and then growing them in controlled environments is biologically very difficult and socially out of the question. The closest that investigators are reported to have come to finding human beings of identical genotypes raised in different environments are the reported cases of identical twins who have been raised apart. Identical twins come from the same fertilized egg, and so are genetically identical. If such a pair could be raised in different environments, then we would have at least two points in the norm of reaction of their genotype. We do not, however, live in a Gilbert and Sullivan operetta world, where children are separated from their parents at birth or exchanged in the cradle and raised in radically different environments. The only published studies of identical twins raised apart that claimed completely random adoptions were those of Cyril Burt, but it has now been revealed that these twins were fictitious. The other published study with a substantial number of twin pairs (43) shows clearly that twins separated at birth are not separated very widely but are kept by sisters, aunts,

The left and right sides of an individual *Drosophila*, showing the difference in the number of sternopleural bristles on each side.

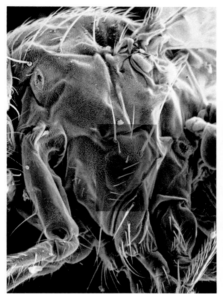

grandmothers, or close friends, usually living in the same region or even in the same village, and often attending school together. We simply do not have the data necessary to plot even two points on a norm-of-reaction curve. Thus, our assessment of the interaction between genes and environment in influencing the development of human beings necessarily depends on historical, anthropological, and social information and on information about the molecular and developmental bases of different traits.

Although no one has ever determined the norm of reaction for human blood groups in relation to, say, temperature or nutrition, we are certain that blood type is insensitive to these variables because of information we have on the molecular mechanism underlying the formation of red blood cell antigens. This confidence is supported, although by no means conclusively, by the fact that people have never been observed to change their blood type in the course of a lifetime. In itself, that stability is not evidence of a constant norm of reaction. Finally, the relation of blood types of children to those of their parents is particularly simple. For example, two parents of blood type M have children only of type M, whereas two type N parents have type N children only. Children of type MN are the only outcome of a mating of an M parent with an N parent.

The phenotype of an individual organism is not completely specified even when its genotype and its developmental environment are given. There is a third contributing cause of variation. Consider the *Drosophila* whose left and right sides are shown in the photographs above. The number of sternopleural bristles, outlined in color, is six on the right side of the fly but ten on the left. Not all flies are "left-sided" in this way—an equal number of flies can be found with fewer bristles on the left than on the right. What is the source of this asymmetry? The two sides of the fly are genetically identical. The fly developed these bristles during the pupal period, when it was adhering by its ventral surface, straight up-and-down, to the inside of a glass culture vessel. No reasonable meaning of the word "environment" would allow us to claim that the left and right sides of the fly developed in different environments. Yet the fly is asymmetrical. The difference between its sides is a consequence of chance events during development. It is *developmental noise*. The formation of a bristle depends upon a bristle-forming cell being present at just the right time in the epidermal layer, the timing and number of cell divisions of the primordial bristle cells, and the migration of daughter cells into the correct layer of the epidermis. Small variations in the concentration and location of molecules within cells will create chance variations in the number of bristle-forming cells that are in the right place at the right time. In *Drosophila,* at least, the variation in bristle number from one side of a fly to another is as great as the variation between individuals, so developmental noise is not a trivial source of phenotypic variation. Wherever cell growth and cell division are involved, we can expect such noise to contribute its effects. The exact placement of hair follicles on our heads, the distribution of small moles on our bodies, a hundred such small details of our morphology are largely under the influence of such random events in development.

One outcome of developmental noise is that differences between individuals may be present at birth, yet not be a consequence of genetic difference. So, for example, I may very well lack the neural connections possessed by Yehudi Menuhin that enable him to be a violin virtuoso while I am a mediocre amateur musician. Moreover, these differences may have already been present when we were newborn infants, yet they may not be a consequence of our different genotypes. The interconnections among the billions of neurons in the brain that arise during development cannot possibly be specified precisely by the genotype, even in a fixed environment. Developmental noise must play a role in the growth of the brain, perhaps a considerable one.

Our discussion of the interaction of causes in determining the variation among organisms suggests a number of models of the developmental process. The first, in which genes are seen as the real determinants of organisms, analogizes development to factory production. The genes are the blueprints and the environmental factors are like raw materials that are processed into the end product, the organism. In this picture, the environmental variables are of a general kind that can be turned into any sort of organism, depending on the genes, just as steel, rubber, glass, plastic, and paint can be converted into trucks, se-

A model of development in which genes are determinative.

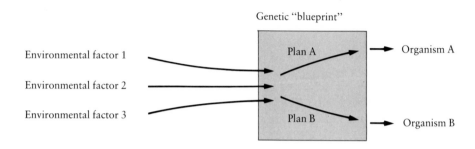

A model of development in which the role of environment is emphasized.

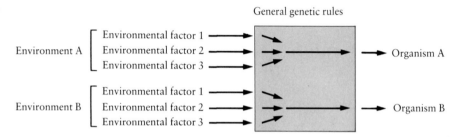

dans, or station wagons, depending upon the particular blueprints being used. A model that emphasizes genetic variation as the dominant source of variation is shown in the first diagram above.

An alternative, complementary metaphor puts the emphasis on the environment as the main determinant of the phenotype: We imagine a building contract that specifies only certain general properties of the building, such as "floors to bear a live-load of 30 pounds per square inch" or "walls with an insulation factor of R19." These are general genetic specifications, but it is the environmental variables that will determine the actual appearance and structure of the organism, just as the choice of materials—wood, steel, or stone—will affect not only the appearance of the building, but also the actual details of its structure. A model that stresses the supremacy of environment in the determination of the phenotype is shown in the second diagram above.

The model that we have built, however, is symmetric with respect to genes and environment; it also includes developmental noise (see the diagram on the next page). Our model is not so much that of a factory as that of an artisan's workshop. The cabinet maker knows what she is to produce, but, as she works, the materials, with their inevitable variation in texture and quality, begin to exert their own influences. The shape of the table legs emerges in part as a response to the influences of the materials. Moreover, because it is handwork,

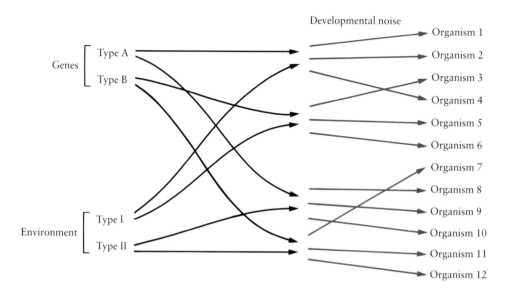

Correct model of development showing the interactions between gene, environment, and developmental noise.

there is some variation from leg to leg and from table to table simply because the worker is not in perfect control of the process. The relation between gene, environment, and organism is not one-to-one but many-to-many. Given the genes and the environment, one cannot predict the organism. Given the organism, we cannot infer its genotype or the environment in which it developed.

For human beings, a model of biology is not enough. Superimposed on the gene and the environment is human self-consciousness, which acts as a vehicle for social interactions to influence individual development. The ways in which men and women carry their bodies, pitch their voices, carry out productive work, and behave in social interactions differs (on the average) between the sexes. Those differences are a consequence of the different socializations of boys and girls, yet they derive historically from some biological differentiation of males and females. The characteristics of "maleness" and "femaleness" develop through interaction between biological differences and social convention. Thus, variation among individual human beings can not be understood from biological principles alone. There are laws of social transformation, laws of which we know very little. It is these social laws that transform a collection of human beings into a human society. But it is also these laws that, reciprocally, govern the development of individual people as affected by social organization. They are the laws of mutual codetermination of the individual and the collective.

Simple Genetic Diversity

3

The red blood cells of all human beings contain hemoglobin, a protein that serves to carry oxygen from the lungs to the innermost recesses of the body. However, all people do not have exactly the same form of this molecule. Most possess the common form, hemoglobin A, but in western and central Africa, in parts of southern India, and in the Arabian peninsula, about one-quarter of the population carries another molecular form, hemoglobin S. People with only hemoglobin S suffer a severe anemia because this molecule forms large crystalline structures inside the red blood cells, causing them to take on a characteristic sickle shape and then to break down. This condition is known as sickle-cell anemia. In populations with hemoglobin S, only about 1%–3% of the people actually suffer from this anemia. About 25%–30% of the people in such populations have both hemoglobin S and hemoglobin A. They suffer only a very mild anemia, and their blood cells rarely deform, but they are significantly more resistant to the severest form of malaria—a form caused by a particular species of malarial parasite, *Plasmodium falciparum*. It is not surprising that the frequency of hemoglobin S is high precisely in those regions of the world where *falciparum* malaria was (and in some places still is) a major disease.

This difference in the form of the hemoglobin molecule is an example of a widespread type of human biological diversity. The characteristics of this type of diversity are:

1. Each person can be classified unambiguously into one of two or more distinct qualitative classes, with no continuum of values between them. Either you have hemoglobin S or you do not. In fact, in addition to hemoglobin A and hemoglobin S, there are more than 300 human hemoglobin variants, but nearly all of these are extremely rare, usually known only from a single person or family. Despite the very large number of variants, each is a clearly distinct *qualitative* variant detectable on the basis of its chemical properties. This kind of variation, called *polymorphism,* is fundamentally different from continuous variation, the kind of variation we observe in such traits as height, weight, form, color, metabolic rate, or behavior, each of which lies on some continuous scale. While we may use categories like "tall" and "short" or "thin" and "fat" as a rough description of continuous variation, such categories are arbitrary and do not really characterize the variation.

2. Qualitative polymorphisms usually, but not always, can be traced directly to alternative forms of some biologically active molecule, as in the case of hemoglobin S and hemoglobin A.

3. These simple polymorphisms are usually unaffected in their primary molecular difference either by alterations in the environment or during the course of development. That is, people born with hemoglobin S will have it throughout

A few normal, disc-shaped red blood cells surrounded by distorted sickle cells in the blood of a person who has sickle-cell anemia.

A space-filling model of the amino acid glycine and the general structural formula of an amino acid.

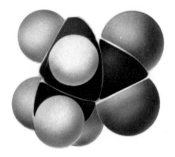

$$R—\overset{\displaystyle H}{\underset{\displaystyle NH_2}{C}}—COOH$$

their entire lives and in any known environment. It is important to understand that this environmental and developmental insensitivity applies only to the forms of the variant molecules themselves, not necessarily to the physiological effects of those molecules. The difference between the fixity of the molecular variation itself and the malleability of its physiological manifestation is clearly exemplified in the case of the different hemoglobin forms.

The developing fetus has a special form of hemoglobin, hemoglobin F, which differs from hemoglobin A in its molecular structure. As the fetus develops, hemoglobin F is gradually replaced by adult hemoglobin. By the time a baby is about six months old, only about 1% of its hemoglobin is of the fetal type. It is the adult hemoglobin that is polymorphic for the A and the S forms; so the difference between carriers of hemoglobin S and those who have only hemoglobin A does not become physiologically apparent until after birth. Moreover, the effect of carrying hemoglobin A or hemoglobin S depends upon the environment after birth. Normally, the red blood cells of a person with both hemoglobin A and hemoglobin S do not sickle and break down, but, if the oxygen tension becomes abnormally low—as it does, for example, at very high altitudes—the crystallization will occur. In addition, the life expectancy of people with and without hemoglobin S depends on their environment. Where *falciparum* malaria is common, hemoglobin S may be a life-saver, but in the United States, for example, it is of no advantage at all. Whether or not hemoglobin S has a physiological effect on its carrier, however, its presence can be detected chemically.

4. The pattern of inheritance of these simple polymorphisms is itself simple. A predictable proportion of the offspring of any particular mating will be of each type. Thus, if both parents have only hemoglobin A, all of their offspring will have only hemoglobin A, and a person with only hemoglobin S must have inherited genes for that form of the molecule from both parents.

Molecular Polymorphism

Hemoglobin is a protein. That is, it consists of long chains of smaller molecules, amino acids. The structure and model of an amino acid are shown at the left above. At one end of each amino acid is an amino group, $-NH_2$, and at the other, an acid group, $-COOH$. Attached to these is another chemical group, symbolized by R, that varies from one amino acid to another. The 20 amino acids commonly found in proteins are listed in the table on the facing page. They fall into three classes, depending on the nature of the R group. Two amino acids, *lysine* and *arginine*, have R groups that are positively charged in solution. Three, *aspartic acid, glutamic acid,* and *histidine,* have R groups that are negatively charged in solution, while the remaining 15 common amino acids have R groups that do not carry any charge.

Classification of amino acids by side groups, with standard abbreviations

Neutral	Neutral	Basic (Positive)	Acidic (Negative)
Alanine (Ala)	Methionine (Met)	Arginine (Arg)	Glutamic acid (Glu)
Asparagine (Asn)	Phenylalanine (Phe)	Lysine (Lys)	Aspartic acid (Asp)
Cysteine (Cys)	Proline (Pro)		Histidine (His)
Glutamine (Gln)	Serine (Ser)		
Glycine (Gly)	Threonine (Thr)		
Isoleucine (Ile)	Tryptophan (Trp)		
Leucine (Leu)	Tryosine (Tyr)		
	Valine (Val)		

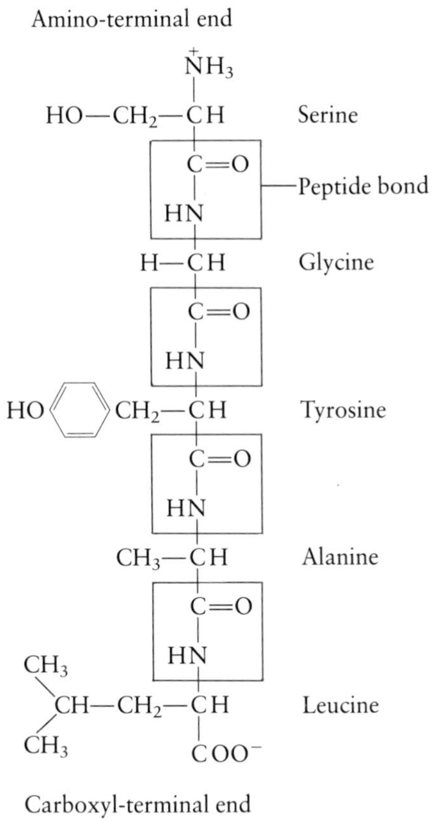

Amino-terminal end

$\overset{+}{N}H_3$

HO—CH$_2$—CH   Serine

C=O   Peptide bond

HN

H—CH   Glycine

C=O

HN

HO⬡CH$_2$—CH   Tyrosine

C=O

HN

CH$_3$—CH   Alanine

C=O

HN

CH$_3$—CH—CH$_2$—CH   Leucine

COO$^-$

Carboxyl-terminal end

Structure of a polypeptide comprising five amino acids, with peptide bonds outlined in color.

Proteins are formed by the end-to-end attachment of amino acids, the –COOH end of one amino acid being joined to the –NH$_2$ group of the next in line, with the elimination of a molecule of water. Such an end-to-end joining, which is called a peptide bond, is shown in the figure at the left. The long amino acid chain, or polypeptide, then folds into a three-dimensional structure held together by bonds between various of the R groups of the different amino acids. The three-dimensional configuration is a unique consequence of the particular order of amino acids in the polypeptide chain, so that the substitution of even one amino acid for another will cause a change in the shape of the three-dimensional folded polypeptide—sometimes a drastic change. An active protein may consist of only a single folded polypeptide or it may be made up of two or more held in a loose chemical association. Sometimes the polypeptide is joined to some other large molecule, such as a carbohydrate chain, to form the final protein.

Hemoglobin is in many ways a typical protein structure. It is shown in the figure on the next page. There are four polypeptides, in two pairs, each of which consists of an α chain and a β chain. The α chain has 141 amino acids; the β chain, 146. If the two chains are compared, one can see some similarities in the order of their amino acids, a consequence of their common evolutionary origin from what was originally a single kind of chain. Each of the four folded polypeptide chains has a pocket in it, into which is inserted a small nonprotein group called *heme,* which contains an iron atom. Such iron atoms are the sites of attachment of the oxygen atoms that hemoglobin carries when it is performing its physiological function. All of the physiological characteristics of the hemoglobin molecule, including the ease with which oxygen is attached and detached, its solubility, the effect upon it of acidity or alkalinity of the blood, and so on, are functions both of the four heme groups and of the four folded polypeptide chains that constitute the molecule. Even single amino acid substitutions can have profound effects on the molecule's properties. Hemoglobin S is

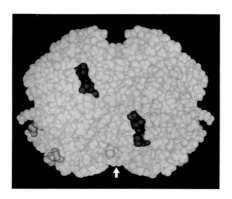

Space-filling model of a hemoglobin molecule. The arrow shows the position on the β chain, β, where valine is substituted for glutamic acid in sickle-cell hemoglobin. Two of the heme molecules are shown as very dark chains near the center.

just such a case. It differs from hemoglobin A only in the substitution of a valine for a glutamic acid at position 6 of the β chain. The three-dimensional view of the hemoglobin molecule (shown at the left) shows the site of this substitution. This single change is responsible for the tendency of hemoglobin S to crystallize within the red blood cells. The three hundred or so known variants of human hemoglobin are all caused by substitutions of single amino acids at various places in the α and β chains. Because there are 287 chain positions in the α and β chains combined, and 20 different common amino acids available to fill each position, there could be $20 \times 287 = 5740$ variants that differ from hemoglobin A by only a single amino acid. So far, only a small fraction of these variants have been discovered, but no doubt many more will appear in time.

Variation in the structure of a protein may have such clear-cut physiological effects that the variation can be detected by the changes it produces in the anatomy or the physiology of its carrier. However, because most such changes will be deleterious to a person's health, we might expect them to be rare in the population. Indeed, such changes are rare: They are examples of genetic diseases rather than of widespread genetic polymorphisms. (For example, a defective form of the enzyme phenylalanine hydroxylase is the basis of phenylketonuria, or PKU, a disease whose chief symptom is early severe mental retardation.) Hemoglobin S is an exception to this rule. Although people who have only hemoglobin S suffer from a disease, sickle-cell anemia, those who have both hemoglobin A and hemoglobin S are protected from malaria. There might, however, be many protein polymorphisms that have no apparent effect on physiology or growth. How might such polymorphisms be detected?

Two methods for the detection of polymorphisms have been widely used. Both depend upon the changes in molecular structure that result when one amino acid is substituted for another in a polypeptide or when the nonpolypeptide part of a protein is changed. The first method detects changes in molecular *shape* and makes use of a natural feature of mammalian physiology, the immune system. The second, electrophoresis, detects changes in molecular *charge* and is purely a chemical procedure for the laboratory.

Blood-Type Polymorphism

When a foreign substance comes in contact with the cells of a vertebrate, a protective reaction occurs. Special proteins, *antibodies,* are produced that fit specifically onto molecules of the foreign substance, the *antigen,* to form an antibody-antigen complex. These complexes are then swept up by the body's white blood cells and broken down into a harmless form. Each different antigenic substance will induce the formation of a protein whose shape is a fairly exact fit to the antigen. The fit is not absolutely specific, so that the same antibody may also combine with other antigens of similar shape. Nevertheless, the fit of an antibody to the antigen that induced it is much better than its fit to other antigens. Virtually any substance can be an antigen. The more complex the shape of the antigen molecule, the more likely it is to induce the formation of antibody molecules with a very specific complementary shape. Precisely how such a bewildering variety of antibodies can be produced on call is not understood.

If red blood cells from a particular person are injected into a rabbit, the rabbit will make antibody molecules specifically to fit the molecular shape of surfaces of the human red blood cells. When these antibody molecules are extracted from the rabbit's serum and mixed on a slide with red blood cells from the same person, an antibody-antigen complex will be formed: Large numbers of the red blood cells will be glued together by the antibody and will precipitate in clumps. If red blood cells from other people are then tested against the antiserum, some samples will clump but others will not. Some people have red blood cells that are similar to those of the person whose cells were used to make the antibody, but some do not. If the entire experiment is repeated using another person's red blood cells, new antibodies will be formed by the rabbit that may not coagulate the cells of the first person. Evidently, there is a polymorphism of the structure of the molecules on the surfaces of red blood cells, a polymorphism that can be detected by making antibodies to the different forms and testing the cells of different people against the different antibodies. It is now known that the structures on the surfaces of red blood cells that are antigenic are not the peptides of the membrane but long chains of sugars of different kinds that are attached to the peptides, forming substances called glycoproteins. Each blood type has a different sequence of sugars in the chain.

The polymorphism of blood types turns out to explain the variable outcome of blood transfusions. Before the turn of the century, it was known that some people were incompatible with others in blood transfusions. A transfusion between an incompatible pair could result in massive coagulation of the transfused blood and the death of the unfortunate recipient. Between 1900 and 1902, Karl Landsteiner discovered that this incompatibility is the consequence of a blood-type polymorphism and that people normally carry antibodies against red blood cells of types other than those to which their own red blood cells belong. The blood types discovered by Landsteiner are shown in the table on the next page, together with the outcome of transfusions between people of the various types. There are four types in this polymorphism, and all people ever examined belong to one of these. People of type A have the A antigen on their red blood cells and the anti-B antibody circulating in their blood serum. If blood from a type B person is transfused into a type A person, the B

Type	Cell Antigen	Antibody	Compatible Transfusions
A	A	Anti-B	Donate to A and AB Receive from A and O
B	B	Anti-A	Donate to B and AB Receive from B and O
AB	AB	None	Donate to AB Receive from all types (universal recipient)
O	None	Anti-A and anti-B	Donate to all types (universal donor) Receive from O

ABO blood types, their antigens, antibodies, and acceptable transfusions

blood cells will be coagulated immediately by anti-B antibody in the recipient's blood serum. In like manner, type B people, who have anti-A antibody, will coagulate blood from a type A donor. People of a third type, type AB, have red blood cells with both kinds of antigen, but they have no antibody. (Obviously, if such people were to have either anti-A or anti-B antibodies, they would coagulate their own blood cells.) Finally, there is type O. People of this blood type have red blood cells that have no antigenic structure, but they have both anti-A and anti-B antibodies circulating in their serum. One consequence of this polymorphism is that people of type AB are universal recipients, since they will not coagulate the blood of any donor. The small amount of antibody that is introduced into them with the donor's blood is so diluted that their own cells are not coagulated by it. Conversely, people of type O can donate blood to people of any type, but they can receive blood only from other people of type O.

Human beings are remarkably polymorphic for this ABO blood-type system. In most European populations, about 45% of the people have type O blood, 35% have type A, 15% have type B, and 5% have type AB. There is some variation from one major region of the world to another, but type O is nearly always the most common and type A the next. Some populations—for example, many American Indian tribes—lack types B and AB entirely.

Since the original discovery of the ABO blood groups in 1900, many other blood-group polymorphisms have been found by antibody-antigen tests. A few of these also create transfusion incompatibilities, and one, the Rh polymorphism, can create a serious incompatibility between a pregnant woman and the fetus she is carrying. Most, however, are without any known chemical or physiological significance. They have been detected entirely by the creation of antisera in laboratory animals and the testing of many people's blood cells against these antisera. Apparently, the human red blood cell carries many different anti-

Blood-type frequencies in the English
white population

System	Type	Frequency
ABO	A	.447
	B	.082
	AB	.034
	O	.437
MNS	MS	.201
	Ms	.093
	MNS	.260
	MNs	.236
	NS	.060
	Ns	.149
Rh	r	.147
	R_1	.535
	R_2	.150
	R_1R_2	.129
	R_0	.022
	r′	.011
	r″	.006
P	P_1	.266
	P_1P_2	.499
	P_2	.234
Secretor	Se^+	.773
	Se^-	.227
Duffy	Fy^a	.177
	Fy^aFy^b	.462
	Fy^b	.301
Kidd	Jk^a	.583
	Jk^aJk^b	.361
	Jk^b	.056
Dombrock	Do^a	.664
	Do	.336
Auberger	Au^a	.857
	Au	.143
Xg	Xg^a	.894
	Xg	.106
Sd	Sd^a	.912
	Sd	.088
Lewis	Le^a	.224
	le^a	.776

Source: Data from R. R. Race and R. Sanger,
Blood Groups in Man, 6th ed. (Blackwell, 1975).

genic sites on its cell membrane, each of which can exist in several alternative forms. The adjoining table lists the frequencies of the different forms for a number of these polymorphisms in the English population. Every person can be classified with respect to each system. Everyone is some ABO type, some MNS type, some Rh type, and so on. If the chance of being of a particular ABO type is independent of one's MNS type, then the chance of belonging to a particular combination of types is simply the product of the frequencies of the types. For example, we expect that the proportion of the English population that is A, MS, P_2 would be $.447 \times .201 \times .234 = .021$. That is, only 2.1% of the English will have this particular combination of ABO, MNS, and P blood types. If we apply this calculation to the entire list in the table, we come to an extraordinary conclusion. It is obvious that the most common combination of blood types will be the combination of the types most common within each polymorphism: It will be A, MNS, R_1, P_1P_2, Se^+, Fy^aFy^b, Jk^a, Do^a, Au^a, Xg^a, Sd^a and le^a. But the frequency of this most common type will be a mere $.447 \times .260 \times .535 \times .499 \times .773 \times .462 \times .583 \times .664 \times .857 \times .894 \times .912 \times .776 = .002$, or less than one-fifth of one percent of the population! Every other type will be even less common. The chance that two people selected at random will both have this most common combination of blood types will be only $(.002)^2$, or about four in a million.

We can carry the calculation further. Instead of asking how likely two people are to belong to the most common type, we could calculate how likely two randomly chosen people are to be identical with respect to the twelve polymorphic blood groups in the table, including all possible combinations. The variety of combinations alone is staggering. Since there are four ABO types, six MNS types, seven Rh types, and so on, there are $4 \times 6 \times 7 \times 3 \times 2 \times 3 \times 3 \times 2 \times 2 \times 2 \times 2 \times 2 = 290{,}304$ distinct combinations of blood types. The chance that any two people will have the same combination of types, irrespective of what that combination is, turns out to be only about three in 10,000.

The small chance that two randomly chosen people are identical in their blood groups takes account only of the twelve blood groups listed for the English white population. There are 50 other blood-group antigen systems known in humans, but these add very little to the distinctions among people, at least among the English, because most people in the population are identical with respect to these nearly monomorphic systems (although some of the systems are polymorphic in other populations). Thus, more than 93% of the English population are of the type Lu^a in the Lutheran blood group and 99.8% are of type K in the Kell blood group. If we were to take account of all the blood groups known, the probability of identity between any two people would be only a little lower. Indeed, about two-thirds of all blood groups are not really polymorphic at all. They correspond to antigens that are known only from a single family ("private" antigens) or that are present in every one in the world except for a rare family ("public" antigens). What is remarkable is that, on the basis of only the twelve antigenic groups, there exists so much biochemical individuality in the English white population.

The HLA Antigens

Beginning about 25 years ago, it was discovered that human lymphocytes, or white blood cells, have antigenic specificities, similar to those of human red blood cells, that are polymorphic. At first it was supposed that only two antigenic series were polymorphic, with several alternative forms each, but recently the degree of variation for the HLA (human lymphocyte antigen) complex has been shown to be truly enormous. There are four antigenic specificities in the HLA system that have been clearly defined: A, B, C, and DR. Among Europeans, 15, 18, 7, and 9 different alternative variants of the four antigens, respectively, are currently known. The frequencies of the alternatives are shown in the table on the facing page. Because no alternative form of any one of the antigens is very common, the degree of individual diversity is immense. It should be noted that the frequencies given in the table are those of the alternative antigenic specificities, but, as for blood-group antigens, each person may carry two alternatives for each specificity. Thus, a person might be A_1A_3, B_8B_{12}, CW_3CW_6, DRW_1DRW_4. The number of combinations is so large (there are 154 for the A antigen alone) that it is clearly impractical to list them all with their frequencies. The alternative variants of the four specificities together give rise to about 25 million distinguishable antigenic combinations.

Enzymes

The second method that has revealed protein polymorphism depends upon the fact that some of the amino acids in a polypeptide are electrically charged in solution. The amino acids lysine and arginine are positively charged because they have an R group that accepts positively charged hydrogen ions from water. The amino acids aspartic acid, glutamic acid, and histidine, by contrast, are negatively charged because they give up hydrogen ions to the surrounding water. Because a protein consists of a chain of amino acids, some of which are positively or negatively charged, the protein as a whole will have a net electrical charge. To a first approximation, the protein is like an electrostatically charged particle. It will move in an electric field in a direction (and with a speed) that depends upon its charge. If the charge of the protein is changed, say by substituting an uncharged amino acid (such as valine) for a charged one (such as glutamic acid), the altered protein will move at a different rate in the electric field. This is the basis of *electrophoresis,* a very common and powerful laboratory method for detecting even small changes in proteins.

Frequencies of different types for HLA antigens *A*, *B*, *C*, and *DR* in samples from three populations

A Antigen	European Caucasoids ($n = 228$)	African Blacks ($n = 102$)	Japanese ($n = 195$)	B Antigen	European Caucasoids ($n = 228$)	African Blacks ($n = 102$)	Japanese ($n = 195$)
A1	.16	.04	.01	B5	.06	.03	.21
A2	.27	.09	.25	B7	.10	.07	.07
A3	.13	.06	.007	B8	.09	.07	.002
A23	.02	.11	—	B12	.17	.13	.07
A24	.09	.02	.37	B13	.03	.02	.008
A25	.02	.04	—	B14	.02	.04	.005
A26	.04	.05	.13	B18	.06	.02	—
A11	.05	—	.07	B27	.05	—	.003
A28	.04	.09	—	B15	.05	.03	.09
Aw29	.06	.06	.002	Bw38	.02	—	.02
Aw30	.04	.22	.005	Bw39	.04	.02	.05
Aw31	.02	.04	.087	B17	.06	.16	.006
Aw32	.03	.02	.005	Bw21	.02	.02	.02
Aw33	.007	.01	.02	Bw22	.04	—	.07
Aw43	—	.04	—	Bw35	.10	.07	.09
Blank	.02	.11	.04	B37	.01	—	.008
				B40	.08	.02	.22
				Bw41	—	.02	—
				Bw42	—	.12	—
				Blank	.04	.18	.08

C Antigen	European Caucasoids ($n = 321$)	African Blacks ($n = 101$)	Japanese ($n = 203$)	DR Antigen	European Caucasoids ($n = 334$)	African Blacks ($n = 77$)	Japanese ($n = 164$)
Cw1	.05	—	.11	DRw1	.06	—	.05
Cw2	.05	.11	.01	DRw2	.11	.09	.17
Cw3	.09	.06	.16	DRw3	.09	.12	—
Cw4	.13	.14	.04	DRw4	.08	.04	.14
Cw5	.08	.01	.01	DRw5	.15	.07	.05
Cw6	.13	.18	.02	DRw6	.09	.10	.07
Blank	.47	.50	.53	DRw7	.16	.07	—
				W1A8	.06	.07	.07
				Blank	.21	.45	.45

Source: H. Harris, *The Principles of Human Biochemical Genetics*, 3rd ed. (North Holland, 1980).

The method of electrophoresis is diagrammed on the next page. A gel is made from starch, agar, or an artificial polymer in order to provide a porous solid medium for the movement of the proteins. The proteins to be examined are placed in holes cut into the gel, which is then immersed in a buffer solution that will hold the chemical composition of the system constant. Next, the two ends of the gel are attached to opposite poles of an electric power supply to create an electric field. When the current is applied, all the protein molecules in the holes begin to move through the microscopic pores of the gel. The positively charged

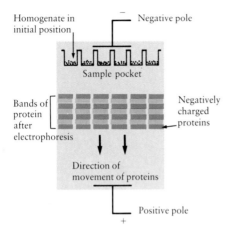

molecules move toward the negative pole, and the negatively charged molecules move in the opposite direction. The molecules of each kind of protein move at a rate that depends upon their net charge and their size. After a few hours, the various proteins will be at different distances from the origin. The problem now is to see the proteins in the gel. For hemoglobin, that is no problem because it is red, and large amounts of the protein from blood can be applied to the gel. To make it even more visible, the gel can be stained with a dye specific to proteins. The figure below shows just such a stained hemoglobin gel. Each band in the gel is the stained protein that has moved from the origin. The first track shows a single band of hemoglobin A. The second track is hemoglobin taken from a person with sickle-cell trait. In each case, the proteins have migrated toward the positive pole, but hemoglobin S has migrated less than hemoglobin A. That is precisely what we would expect, from our knowledge of the chemical difference between these two molecules. Hemoglobin S differs from hemoglobin A by hav-

Diagram (above) of gel electrophoresis of proteins. Electrophoretic patterns (right) of the hemoglobin of someone who has sickle-cell trait and that of someone who has sickle-cell anemia, compared with normal hemoglobin.

Phenotype	Genotype	Hemoglobin electrophoretic pattern Origin ⟶ +	Hemoglobin types present
Normal	AA		A
Sickle-cell trait	AS		S and A
Sickle-cell anemia	SS		S

ing a valine substituted for a glutamic acid at position 6 of the β chain. Because it has lost a negatively charged amino acid, its net charge is less negative than hemoglobin A, and it will move more slowly toward the positive pole.

As it has been described, gel electrophoresis should only be able to detect a fraction of all amino acid substitutions. It should not, for example, be able to distinguish hemoglobin E (which is a variant with a substitution of lysine for glutamic acid at position 26 of the β chain) from hemoglobin Agenogi (which has the identical substitution at position 90 of the β chain instead). Both have a positive charge substituted for a negative charge; indeed, exactly the same

Electrophoretic patterns of hemoglobins with different amino acid substitutions. The same chemical substitutions at different locations on the molecule are detectably different in electrophoresis. For each hemoglobin, two samples are run in adjacent tracks to show how repeatable the electrophoretic method is when the same molecule is run separately.

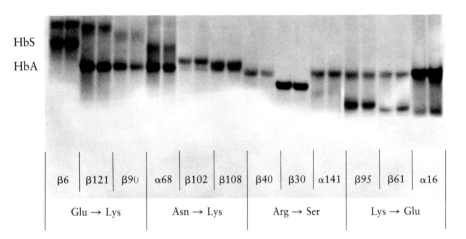

HbS

HbA

β6	β121	β90	α68	β102	β108	β40	β30	α141	β95	β61	α16
Glu → Lys			Asn → Lys			Arg → Ser			Lys → Glu		

amino acids are involved, only in different positions in the molecule. The figure above shows, however, that these two hemoglobins are, in fact, separable by electrophoresis, as are other proteins with "identical" substitutions. This is because the degree of gain or loss of hydrogen ions by a charged amino acid in a protein depends not on that amino acid alone but on the other amino acids in its immediate neighborhood in the molecule. So, even the substitution of one neutral amino acid for another—say, methionine for valine, as in hemoglobin Kalu—will cause neighboring charged amino acids to alter their net charge. At present, it is estimated that 85%–90% of all amino acid substitutions in proteins can be detected electrophoretically.

Most proteins are not present in sufficient concentration in tissues that we can detect them on gels by means of a protein stain. Red blood cells contain many enzymes whose function is to catalyze the various metabolic reactions of a living cell. Their concentrations are so low, however, that they cannot be seen in the protein-stained gel shown above. To visualize them, we can make use of their catalytic properties. Each enzyme works on a particular substrate material, making a specific chemical change in it and converting it into another molecule. In the process, the enzyme molecule is not used up but is released to work again on yet another substrate molecule. So, for example, the enzyme glucose-6-phosphate dehydrogenase (G6PD) converts glucose-6-phosphate into 6-phosphogluconate by removal of a specific hydrogen. By providing the appropriate intermediate compounds and a dye, we can cause the dye to become colored and precipitated whenever such a hydrogen is made available. The consequence will be that a colored band will appear in the gel at the place occupied by the enzyme. It is not the enzyme that is being stained by the dye; in a sense, it is the enzyme's activity.

The electrophoretic technique can be applied to an immense range of enzymes. All that is required is a coupled dye reaction appropriate for each. No purification of the enzymes is needed. Blood cells or bits of tissue can be macerated, the solid debris centrifuged away, and the total tissue extract subjected to electrophoresis. Specific dye reactions and substrates identify each enzyme on the gel. More than 100 enzymes have now been studied in this way in human populations. Of these, 25% have been found to be polymorphic within populations, with more than one alternative form repeatedly found in the same population. Occasional rare variants are found in those enzymes that are not polymorphic, and these too contribute to the biochemical diversity among individual people.

The table below lists the fifteen most polymorphic enzymes in the English population. As for blood groups and HLA types, each person may carry one or two variants of each molecule.

Frequencies of various enzyme variants found in the English population	Enzyme	Form					
		1	2	3	1/2	2/3	1/3
	Red-cell acid phosphatase	.13	.36	0	.43	.05	.03
	Phosphoglucomutase 1	.59	.06	—	.35	—	—
	Phosphoglucomutase 3	.55	.07	—	.38	—	—
	Placental alkaline phosphatase	.41	.07	.01	.35	.05	.12
	Peptidase A	.58	.06	—	.36	—	—
	Adenylate kinase	.90	.01	—	.09	—	—
	Adenosine deaminase	.88	.01	—	.11	—	—
	Alcohol dehydrogenase 2	.94	—	—	.06	—	—
	Alcohol dehydrogenase 3	.36	.16	—	.48	—	—
	Glutamate-pyruvate transaminase	.25	.25	—	.50	—	—
	Esterase D	.82	.01	—	.17		
	Malic enzyme	.48	.09	—	.43	—	—
	Phosphoglycolate phosphatase	.68	.03		.29		
	Glyoxylase I	.30	.21		.49		
	Diaphorase 3	.58	.05		.36	—	—

Source: H. Harris, *The Principles of Human Biochemical Genetics,* 3rd ed. (North-Holland, 1980).

Are Humans Unusually Polymorphic?   If anything, human beings and other mammals are somewhat less biochemically polymorphic than other organisms. During the last dozen years, biologists have studied biochemical variation in many different organisms in order to determine how much genetic variation is available for the evolution of species and how much genetic change has already occurred between species. Electrophoresis makes this task especially easy. Any bit of tissue (or, in the case of small organisms like insects, the entire creature) can be ground up, put on an electro-

Genetic variation in natural populations of some major groups of animals and plants	Organisms	Number of Species	Average Number of Loci per Species	Average Polymorphism	Average Heterozygosity
	Invertebrates:				
	Drosophila	28	24	.529	.150
	Wasps	6	15	.243	.062
	Other insects	4	18	.531	.151
	Marine invertebrates	14	23	.439	.124
	Land snails	5	18	.437	.150
	Vertebrates:				
	Fishes	14	21	.306	.078
	Amphibians	11	22	.336	.082
	Reptiles	9	21	.231	.047
	Birds	4	19	.145	.042
	Mammals	30	28	.206	.051
	Plants:				
	Self-pollinating	12	15	.231	.033
	Outcrossing	5	17	.344	.078
	Overall averages:				
	Invertebrates	57	22	.469	.134
	Vertebrates	68	24	.247	.060
	Plants	17	16	.264	.046

Source: F.J. Ayala and J.A. Kiger, *Modern Genetics* (Benjamin-Cummings, 1980).

phoretic gel, and classified according to its molecular phenotype. This has been done for plants and animals, vertebrates and invertebrates, algae, fungi, and bacteria. The result is rather consistent. About one-third of all the kinds of proteins examined turn out to be polymorphic. In humans, it is only about one-quarter. The table above shows the proportion of polymorphic enzymes in a number of different plants and animals.

The number of different forms of a polymorphic enzyme is often much greater in other animals than it is in humans. The table on the facing page shows that, for fifteen polymorphic enzymes in the English population, thirteen have only two alternative molecular forms while two have three alternatives. In the fruit fly *Drosophila,* however, the typical polymorphic enzyme exists in populations in four or five alternative forms. In *Drosophila pseudoobscura* from western North America, 27 electrophoretically distinct variants of xanthine dehydrogenase were discovered in only 146 independent lines, and 26 distinct variants of an esterase enzyme were found in only 106 lines. Only the HLA system in humans rivals this degree of molecular diversity. So human beings are a bit conservative in their molecular variation, but not so much as to make them stand out. In this respect, if in no other, they are typical mammals.

Biochemical Individuality

In addition to the great diversity of individual people that arises from the major biochemical polymorphisms, there is an immense variety of rare variants even among molecules that are not truly polymorphic. There are 32 "public" and "private" blood groups where variation is known only in a single family. About 25 million different HLA types can be produced from the known different forms of the HLA antigens. For many of those enzymes found to be monomorphic in samples of moderate size, an occasional electrophoretic variant has been known to turn up. H. Harris, D. A. Hopkinson, and E. B. Robson have estimated that 1.75 people per thousand carry a rare variant of some enzyme. Hemoglobin alone is known to have more than 300 rare single amino acid variants. We do not know how many different kinds of enzymes and other proteins the human body is made of. Certainly 10,000 is a conservative estimate. If each one of those proteins had, on the average, some variant that occurred in only one out of a thousand people, the chance that any given person would be completely free of *all* rare variants would be $(1 - .001)^{10,000} = .000045$, or about five in a million. So, each one of us almost certainly differs from all the unrelated people around us by at least one rare protein variant. If we ignore these rare variants and turn back to the common polymorphisms, we reach a more impressive conclusion: Taking into account the probability of identity for blood groups, major HLA antigens, and polymorphisms for the enzymes and adding to the list several other major polymorphisms, we arrive at the total shown in the table at the left. The data given are for Europeans because they have been most studied, but what information has been gathered for Asians and Africans gives the same picture. Even if this book were vastly more successful than in the wildest dreams of its author and publisher, no two of its readers would share the same major biochemical polymorphism. Indeed, no two unrelated humans who have ever lived or ever will live are likely to be identical even for the handful of commonest molecular polymorphisms. Except for identical twins, we are biochemically unique.

Probability of identity for various polymorphisms for two randomly chosen Europeans

Polymorphism	Probability of Identity
Blood groups	.00029
HLA antigens	.000049
Enzymes	.000037
Haptoglobins	.39
γ-Globulin light chain	.85
β-Lipoproteins	.48
Total (product of the above)	.00000000000008 = 8 per 100 million million

The Genetic Basis of Simple Polymorphisms

4

The fact that molecular polymorphisms are apparently invariant throughout a person's lifespan and independent of any observable environmental or cultural influence suggests that they are simply and directly inherited. Indeed, we can use blood-group polymorphisms to develop the laws of simple inheritance even more easily than Mendel used his famous garden peas.

The first observation about the inheritance of blood-group patterns is that there is indeed some biological inheritance of the traits. If both parents are of blood type M, all their children will be of type M. If both parents are of type N, all of their children will be of type N. Blood types M and N are traits that breed true. Two parents who are type A for the ABO system will never have any offspring with the B antigen, and vice versa. Because these family patterns are uninfluenced by any known environmental variation, we must assume that the resemblance between parents and offspring is genetic.

The second observation is that sex is irrelevant. If a male of type M mates with a female of type N, all of their offspring will be of type MN. But the same outcome will occur if the male parent is N and the female is M. Thus, we can conclude that both parents contribute equally to the blood types of their offspring. It is apparently unimportant for the determination of a child's blood type that the egg and sperm that joined together at conception differed in volume by a factor of 200,000, or that the fetus developed in its mother's uterus, or that the child was suckled by her.

The third observation, and one that provides the basic understanding of the simple rules of inheritance, comes from the mating of a person whose parents were of different blood types. Let us suppose that a woman is of blood type AB. Investigation reveals that one of her parents was of type A and the other of type B. Clearly, she has received whatever the determinants of antigenic specificity may be from both of her parents. If this AB woman has several children by a man who is type O—that is, who has no antigenic specificity in the ABO system—each of their children will turn out to be either of type A or of type B. Such a couple will never conceive any AB children or any O children, only A children and B children. Symbolically,

$$AB \times O \quad \text{parents}$$
$$\downarrow$$
$$A \text{ or } B \quad \text{children}$$

How is this curious result to be explained? Since both parents contribute equally to their offspring and since the father is of a blood type with no antigenic specificity, the children clearly must have inherited this lack of antigen from their father. From their mother, each child has inherited one of the two antigenic specificities that she carried, but never both. So, the determinant of

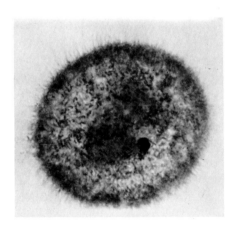

A human ovum surrounded by sperm. One sperm will succeed in penetrating the egg membrane and fertilizing the egg. Changes will then occur in the membrane to exclude the remaining sperm.

Albino parents and their albino daughter.

antigen A and the determinant of antigen B, which came together when the mother of the family was conceived, *came apart again when she had children.* Symbolically,

$$A \times B \quad \text{grandparents}$$
$$\downarrow$$
$$\text{AB mother} \times \text{O father}$$
$$\downarrow$$
$$\text{AO or BO children}$$

This is the fundamental property of inheritance: Determinants of traits that come together at a person's conception come apart again when that person produces sperm or eggs. It is important to note that the mother in our example is of a mixed physical type, an amalgam of the physical traits of her parents. But that mixture at the physiological level does not mean that her hereditary determinants have lost their individuality. On the contrary, they have remained discrete and have become separated again in the formation of her reproductive cells. It is as if, having mixed red and white paint to produce pink, I were able to sort out the molecules in a sample of the pink paint to produce red paint and white paint again. The determinants of heredity are some sort of quanta, discrete particles that maintain their individuality in the process of inheritance, despite the fact that, when they come together at conception, their physiological effects may be blended together. Individual people are only the temporary carriers of mixtures of determinants. Each person is like a pointillistic painting, made up of discrete dots of different colors, but blended into shades when seen from a distance.

These observations are summed up in the fundamental law of segregation discovered by Gregor Mendel in 1865: For any simple trait, an organism receives a determinant from each parent at conception, but these determinants *segregate* (separate again) when the organism produces *gametes* (sperms or eggs). In modern terminology, these determinants are *genes* and the alternative forms of each gene are *alleles.* Each of us carries two genes for each trait, one inherited from each parent. Each of us will, in turn, pass on copies of one of these two genes to each of our children. Which of the two genes each child receives is a matter of chance.

The genes inherited from the two parents may be of the same allelic form, in which case the person who has inherited the genes is called a *homozygote.* If the genes are of different allelic forms, the person is a *heterozygote.* A person with the MN blood type is a heterozygote, carrying an allele that specifies the M antigen and one that specifies the N antigen. An N person, on the other hand, since he or she is carrying two genes, must have inherited the same kind of allele, N, from both parents. Such a person is a homozygote, *NN*. The mating of a person of blood type MN with a person of blood type N will, in genetic terms, be the mating *MN* × *NN*, and, following the rule that every child receives one gene from each parent, there will be two sorts of offspring:

$$MN \times NN$$
$$\downarrow$$

| *MN* or *NN* | Genetic composition |
| MN   N | Blood type |

Were the parents both heterozygotes (both *MN*), the result would be

$$MN \times MN$$
$$\downarrow$$

| *MM* or *MN* or *NN* | Genetic composition |
| M   MN   N | Blood type |

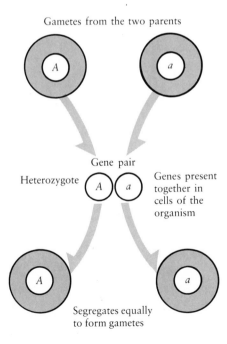

Gametes from the two parents

Gene pair

Heterozygote

Genes present together in cells of the organism

Segregates equally to form gametes

General scheme of the union and segregation of genes in successive generations.

That is, not only would there be a reproduction of heterozygotes *(MN)*, but segregation would result in the production of homozygotes as well. This is perhaps the most important consequence of the law of segregation of genes. Even in the simplest kind of inheritance, like does not simply beget like. When two heterozygotes mate, they also produce homozygotes. Thus, although the two parents are identical, their offspring are heterogeneous. Moreover, because it is a matter of chance which of the particular possible combinations of alleles will come together in a given child, two heterozygous parents with, say, three children could easily produce no children like themselves at all. The reader can verify that the chance of such an event is $(1/2)^3 = 1/8$.

The genetic interpretation of the observed variation in molecular types clarifies why a person may have none, one, or two of the antigens belonging to a given group, but never more than two. Each variant molecular type within a group is the consequence of a different allele of the same gene. Every person either carries only one sort of allele or is heterozygous for two alleles. If the allele results in no antigenic molecule, then a homozygote for that allele will have no antigenic specificity. An individual homozygous for an allele that *does* specify an antigen molecule will have one such specificity, while a heterozygote for two alleles that specify different antigen variants will have both antigens. Some of the complexity of the relation between the genes and the variation in molecular types that they determine is illustrated by the ABO blood group.

According to the law of segregation, we should expect two AB parents to have three sorts of offspring: A, B, and AB. To distinguish the genes from the blood types they determine, we will use the symbols I^A, I^B, and i to denote the gene alleles that specify the A antigen, the B antigen, and no antigen (O). Thus, the A and B offspring from the mating are homozygous—I^AI^A and I^BI^B. Suppose now that one of the homozygous offspring—say, one with type A blood—were to mate with a type O person, whose genetic formula must be ii. The consequence of the mating is $I^AI^A \times ii \rightarrow I^Ai$, a heterozygote whose blood is nevertheless of type A. There are then two sorts of type A people: homozygotes, I^AI^A, and heterozygotes, I^Ai. The only difference between them is in the blood types of the offspring they will produce. For example, a mating between two type A homozygotes (I^AI^A) will produce only children with type A blood, but a mating of two type A heterozygotes (I^Ai) will, in accord with the law of segregation, produce some homozygotes *(ii)* whose blood type is O. Once again, like begets unlike.

Because the same overt difference may have different genetic bases, it is essential to distinguish genotype from phenotype. Two different genotypes, I^AI^A and I^Ai, have the same phenotype, blood type A. There are no general rules that will allow us to predict the relation between genotype and phenotype. Whether a heterozygote will have a unique phenotype or will look just like a homozygote and whether all homozygotes will be distinguishable from each other will depend upon the details of the translation of gene into organism.

Chromosomes as the Carriers of Genes

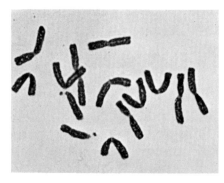

Chromosomes of the plant *Hepatica.* Of the fourteen chromosomes, seven were inherited from one parent and seven from the other, making seven pairs in all.

The characteristics of genes—that each parent contributes equally one of each kind in sperm or egg and that they come together only temporarily in individual people and then separate again at the formation of the next generation of sperm and eggs—are precisely the characteristics of certain cellular bodies, the chromosomes. These are elongated, rod-shaped bodies in the nuclei of cells (see the photograph at the left). The chromosomes in the sperm are the same in number and size as those in the egg, despite the enormous difference in size between the two kinds of cells. The number of chromosomes in a person's body cells is twice what it is in his or her gametes. In fact, in the body cells that divide to produce the gametes, the chromosomes can be observed to be grouped in pairs (see the photograph on the facing page). As the formation of gametes proceeds, one member of each pair passes into each gamete. Occasionally, a small segment of a chromosome is lost. When this occurs, particular genes are found to be missing. In this way, specific genes can be localized to specific segments of a chromosome. Genes are linearly arranged along the chromosomes, and the behavior of genes in inheritance is simply a reflection of the way in which chromosomes separate and rejoin from generation to generation. The choreography of this chromosomal dance has important consequences for genetic diversity.

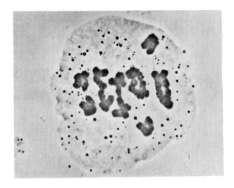

The six pairs of chromosomes in the plant *Tradescantia* beginning to separate into two groups of six each in gamete formation.

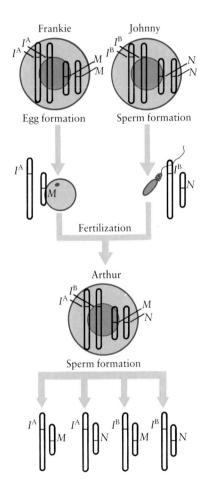

The inheritance of genes for two different traits on two different chromosome pairs and the recombination of those genes in gamete formation.

The genes of humans are distributed along 23 pairs of chromosomes. One member of each pair of chromosomes has been contributed by each of the individual's parents. In the process of gamete formation in the gonads, the chromosomes physically pair and line up, as in the photograph at the left, and then begin to segregate into separate gametic cells. The members of each pair of chromosomes undergo this movement independently of the members of every other pair. As a result, each of a person's gametes should receive some chromosomes inherited from the person's father and some from the person's mother. The gamete will then contain combinations of chromosomes that did not exist in the original two gametes that joined to make that person. But this recombination of chromosomes must lead to recombination of genes as well.

Consider Frankie and Johnny. Frankie is a type A homozygote, $I^A I^A$, and type M; so her eggs are all $I^A M$. Johnny is a type B homozygote, $I^B I^B$, and type N; so his sperm are all $I^B N$. Moreover, it is known that the gene for the ABO antigen system is carried by a different one of the 23 chromosome pairs than is the gene for the MN system. Frankie's and Johnny's child, Arthur, will then have the chromosome composition shown in the diagram at the left, and his genotype is $I^A I^B$, MN. When Arthur produces sperm, Frankie's and Johnny's chromosomes will assort to these sperm independently: some of them will be $I^A M$ and $I^B N$, like the parental combination, but others will be $I^A N$ and $I^B M$, which are quite new sorts of gametes. Recombination of chromosomes has resulted in increased genetic diversity.

If gamete formation consisted only in the assortment of chromosomes, then different genes on the same chromosome pair could never recombine. Gamete formation entails another process, however—one in which there is an exchange of parts between the chromosomes of a pair—that will allow new combinations to appear in gametes even when two different systems are associated with the same pair of chromosomes. The recombination of parts is diagrammed on the next page. It occurs roughly once on each chromosome pair during the formation of each gamete. The place along the chromosome at which the exchange occurs varies from one gamete-forming cell to another. Genes that are very close together on the chromosome, such as *J* and *K* in the figure, will almost never recombine because it is very unlikely that the exchange position will fall exactly between them. On the other hand, *J* and *L* will recombine quite often because they are so far apart that most of the positions of exchange will separate them. For example, the gene for the S blood group is so close to the gene for the MN system on human chromosome pair number 2 that they never recombine with each other. If Frankie's genotype had been *MMSS* and Johnny's *NNss*, so that Arthur's were *MNSs*, virtually all of Arthur's gametes would have been *MS* and *Ns* and none would have been *Ms* or *NS*. On the other hand, the gene for the

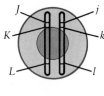

Chromosomes double

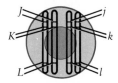

Then some strands exchange parts

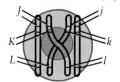

and then are distributed into gametes

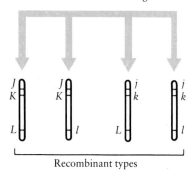

Recombinant types

The recombination of genes that are carried on the same chromosome pair.

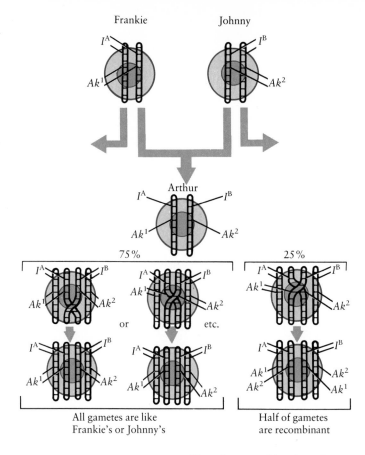

How the recombination of genes separated by one-quarter of the length of the chromosome produces new combinations in one-eighth of the gametes.

adenylate kinase enzyme, Ak, is about one-quarter of the way down its chromosome from the gene for the ABO blood group. If Frankie and Johnny were $I^A I^A$, $Ak^1 Ak^1$ and $I^B I^B$, $Ak^2 Ak^2$, Arthur would produce not only the parental combinations I^A, Ak^1 and I^B, Ak^2 in his sperm but also one-eighth I^A, Ak^2 and I^B, Ak^1 recombinant gametes, as shown above.

Haplotypes

There are a number of systems of genes that control similar functions and that are very close to each other on the chromosomes so that they seldom recombine. Such a very close arrangement of genes with related functions may be a trace of their origin, one from another, in evolution. Several of the polymorphisms already mentioned are of this sort, in addition to the genes for the MN and the S blood groups. The Rh blood group polymorphism has three distinct antigenic specificities, C, D, and E, each of which is probably determined by a separate gene, each with several allelic forms. The HLA system, with its four separate antigenic series, A, B, C, and DR, is a cluster of four very close genes, each with many allelic forms. In each of these cases, the genes are so close together on the chromosome that no one has ever observed a recombination between them, although different combinations exist in the population. So, a man whose own chromosomes are *MS* and *Ns* has never been observed to produce *Ms* or *NS* sperm, as judged by his offspring. Yet all four combinations exist in the species in high numbers; so we must assume that there is, in fact, some recombination and that, over the long period of human biological history, the mixture of all four chromosomal types has become established. Such chromosomal types— combinations of several different genes that are seldom observed to recombine—are called *haplotypes*. A haplotype is transmitted as a unit in heredity. For example, a woman whose HLA constitution is A_2A_3, B_7B_8, $C_{W5}C_{W6}$, $DR_{W2}DR_{W5}$ might have any one of $2^4 = 16$ haplotype chromosomal constitutions. She might have haplotypes $A_2B_7C_{W5}DR_{W2}$ and $A_3B_8C_{W6}DR_{W5}$, or $A_3B_7C_{W5}DR_{W2}$ and $A_2B_8C_{W6}DR_{W5}$, or any of 14 other possible combinations, depending upon which two haplotypes she inherited from her parents: Whatever two haplotypes she was formed from, these are virtually the only two that she will pass on to her children (although about 1% of her eggs will be recombinant forms).

If you look back at the table of HLA types in Chapter 3 (p. 37), you will see that the four HLA genes, *A*, *B*, *C*, and *DR*, have 15, 18, 7, and 9 allelic forms, respectively. The total number of possible haplotypes is then $15 \times 18 \times 7 \times 9 = 17,010$. Most of these have never been observed, partly because the number of people who have been tested is not yet sufficiently large. The haplotype $A_{W33}B_{37}C_{W1}DR_{W1}$ has an expected frequency of only 2 in 10 million; so it would be easy to miss. Some haplotypes undoubtedly do not actually exist, either because they have been lost to the species or because they have perhaps not yet been formed. The entire human population was probably no greater than 5 million people as recently as 10,000 years ago, or only about 400 generations in the past. For most of human history, there have been too few people to include carriers of all the rare haplotypes of the HLA system alone, not to mention carriers of combinations of those haplotypes with those of the Rh and MNS systems. To complicate the picture further, more variants of the HLA genes are discovered each year, and the number of known possible haplotypes therefore continues to grow.

What Genes Are Made Of

The material substance of genes must have two characteristics. First, it must be capable of being duplicated over and over with essentially perfect fidelity: A single set of chromosomes from father and mother come together in the fertilized egg cell. That cell then divides to produce a body with millions of cells— and further hundreds of eggs or millions of sperm cells in the gonads of the

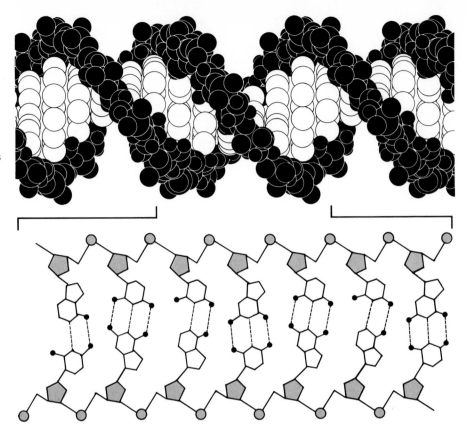

Space-filling model of a molecule of DNA. The black, twisted double helix is a pair of alternating sugar-phosphate chains. The white structures are the atoms of the base pairs connecting the two helical chains. Shown below the model is a short unwound section of DNA.

adult. All of these cells contain copies of the original chromosomes and genes. Second, the genetic substance must be capable of existing in a very large number of different forms corresponding to the thousands of different genes in the set.

Chromosomes consist of two kinds of molecules, protein and DNA (deoxyribonucleic acid), both of which have the needed characteristics. For a long time, it was assumed that the protein was, in fact, the genetic substance: The long chains of amino acids, with any one of 20 amino acids possible in each position, clearly provide the necessary variety needed for the genes and, at the same time, suggest the linear arrangement of the genes on the chromosomes. Moreover, machinery for the synthesis of proteins clearly existed in cells; so it was reasonable to assume that a genetic protein could serve as a template for building up copies of itself using that machinery. About 30 years ago, however, it became clear that it is in fact the DNA that is the genetic material. The evidence came from many sources, but it was mostly based on experiments that showed that purified DNA was capable of carrying hereditary information from one cell to another, whereas protein was not. For example, bacterial cells of one form can be hereditarily transformed into a second type by treating them with purified DNA from cells of that second type. Treatment with protein is ineffective.

Like protein, DNA has a molecular structure that makes possible both accurate self-replication and enormous diversity of form. The structure of DNA, the

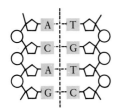

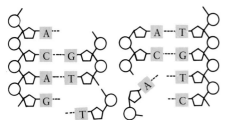

A new DNA molecule is made by using the two complementary halves of the old molecule as templates. The left half-molecule has the sequence A-C-A-G, which acts as a template on which to build a new half-molecule of the complementary sequence T-G-T-C. At the same time, the right half-molecule T-G-T-C acts as a template for a new half-molecule A-C-A-G. This results in two new double molecules, identical with each other and with the original DNA.

famous double helix of Watson and Crick, is shown on the facing page. The molecule is a long helically twisted ladder, the side rails of which are made up of alternating sugar (deoxyribose) and phosphoric acid groups. The rungs of the ladder are made up of pairs of bases. There are only four sorts of bases in DNA—adenine, thymine, guanine, and cytosine—and their spatial structure is such that, if one half-rung on the ladder is adenine (A), the other half-rung must be thymine (T) or else the rung will not fit between the side rails. In like manner, cytosine (C) must always be paired with guanine (G). Choosing one of the deoxyribose-and-phosphate side rails and reading the names of the bases attached to it in order, we might find the following sequence:

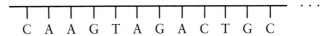

C A A G T A G A C T G C

We are then certain that the other half of the molecule, beginning at the same end, will read:

G T T C A T C T G A C G

since C must always be paired with G and A must always be paired with T. The final element (and a vital one) in the structure of DNA is that the chemical bonds holding the two half-helices together—the bonds between G and C and those between T and A in the middle of the rungs—are hydrogen bonds. These bonds are very weak, and they allow the two halves of the helix to come apart easily, like the halves of a zipper.

The complimentary structure of two half-molecules is the feature that makes the orderly self-replication of the genes possible. At replication, the two half-molecules come apart and new halves are built up on each of them. Because of the spatial restrictions, the new half will always be the perfect complement of the old half. As a consequence, two double helices are built up that are exact copies of the original. One has an old right-hand side and a new left-hand side, the other an old left-hand side and a new right-hand side, as shown in the diagram at the left.

The linear arrangement of the base pairs along the molecule allows for the necessary variety. Suppose that a gene consisted of a DNA molecule 500 base pairs long. Since there are four possible bases at each position along each half-molecule, there are 4^{500} different possible arrangements for each, vastly more than the number of atoms in the universe, not to mention the number of different kinds of genes we require.

How Genes Work

5

We say there is "a gene for the β chain of hemoglobin" and "a gene for the enzyme glucose-6-phosphate dehydrogenase." We also say that one allele of the β-chain gene causes glutamic acid to be in position 6 of the β chain of hemoglobin, while a different allele of the gene changes this to valine, producing hemoglobin S. Both of these statements about genes come down to saying that genes contain the information specifying the sequence of amino acids in a protein. After all, in one sense the difference between hemoglobin and glucose-6-phosphate dehydrogenase is of the same nature as the difference between hemoglobin A and hemoglobin S. It is a question of amino acid sequence. Change a single amino acid and you change hemoglobin A into hemoglobin S. Change enough amino acids and you change hemoglobin into G6PD.

The problem then is to explain how the information in a sequence of bases in DNA is transformed into a sequence of amino acids in proteins. Specifically, how does the cell go from CAAGTAGACTGC . . . to valine-histidine-leucine-threonine- or from the gene for the β chain of hemoglobin to the β chain itself.

Clearly, the information in DNA cannot be converted base by base. There are only four different bases, A, G, T, and C, while there are 20 different amino acids to choose from at each position in the protein. Bases taken two at a time are still insufficient, since there are only $4^2 = 16$ possible couplets. Triplets of sequential bases would be quite sufficient, however, since there are $4^3 = 64$ of them, but then we must suppose that several different triplets can code for the same amino acid. That is, in fact, the case. The protein-synthesizing machinery of the cell reads the string of bases on one of the half-molecules of DNA in groups of three and then translates each group of three into a specific amino acid which is then incorporated into the protein being manufactured. In the example given above, the translation is:

C A A   GTA   GAC   TGC ...
valine-histidine-leucine-threonine-

and so on. These are the first four amino acids in the β.chain of hemoglobin.

The details of the protein-production machinery are shown on the next page. The flow is remarkably like that of an assembly line in a factory. The steps are:

1. A complementary strand of ribonucleic acid (RNA) is built up along the exposed strand of a half-molecule of DNA. The rules of base pairing that apply to the replication of DNA apply also to the synthesis of RNA, except that, in RNA, thymine (T) is replaced by another base, uracil (U). Thus, the RNA is a kind of negative impression of the DNA mold. Individual RNA nucleotides (bases with their attached sugar and phosphoric acid groups) pair with the appropriate bases along the DNA template. These nucleotides are then linked in a long chain by a special enzyme that causes chemical bonds to form between the

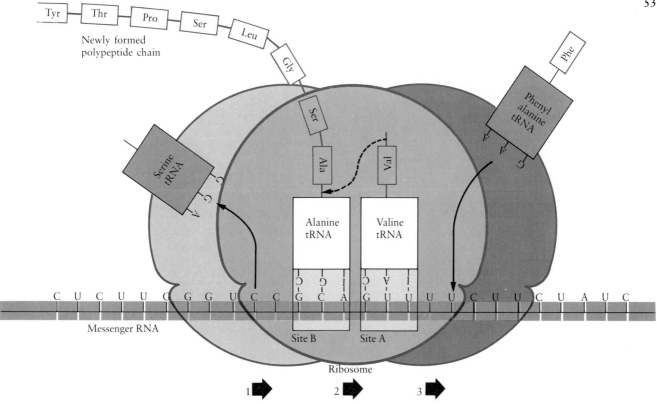

Newly formed
polypeptide chain

Messenger RNA

Site B   Site A

Ribosome

1 ➤   2 ➤   3 ➤

The production of a polypeptide chain by the ribosomal machinery. Note that the triplet code on the valine and alanine tRNAs includes a fifth form of nucleotide, inosine (I), which can be complementary to U, C, or A in the messenger RNA.

sugar group of one nucleotide and the phosphoric acid group of the next. The RNA chain complementary to the DNA given on the facing page would be:

GUU CAU CUG ACG

2. The long, newly made RNA copy falls off the DNA, making room for yet another RNA copy to be made. Once completed, each RNA copy passes out of the nucleus to join the rest of the protein-production machinery of the cell. The making of the RNA copy—aptly called *messenger RNA*—is necessary because the DNA in the chromosomes is isolated, inside the nucleus, from the rest of the machinery of the cell. In addition, many message copies can be made rapidly and exported to the protein-production machinery, without having to wait for that machinery to grind out its product. The process of making the messenger RNA, which is called transcription, can be likened to the workings of a magnetic tape copying machine, churning out working tapes (messenger RNA) from a master copy (DNA).

3. The transcribed RNA copy must now be translated into a sequence of amino acids for the protein. This is done by a dictionary of small folded molecules called *transfer RNAs* (tRNAs) that have built into them the correspondence between RNA triplet and amino acid. At one end of each folded tRNA molecule is a triplet of nucleotides that is complementary to one or another of the messenger RNA triplets. At the other end of the molecule an amino acid is

attached, the amino acid that is to correspond to the RNA triplet being translated. Since there are 64 different RNA triplets to be translated but only 20 different amino acids to be coded for, several different RNA triplets will be translated as the same amino acid. This can occur because some amino acids have several different tRNAs that translate them, and, conversely, some tRNAs can match more than one kind of RNA triplet in the message.

4. The front end of the messenger RNA feeds into one of the cell's many ribosomes, small bodies that function as the cell's protein-synthesizing machinery. As the first code triplet of the message enters the ribosome, it is activated to receive a transfer RNA. By random molecular motion, an appropriate complementary transfer RNA attaches to the activated triplet. The message tape then moves a frame forward into the ribosome machinery, and so the next code triplet is activated. When its transfer RNA attaches to it, two things happen: The amino acid on the second transfer RNA attaches to the amino acid on the first transfer RNA. At the same time, the first amino acid is released from its transfer RNA. Now the tape moves again; yet a new transfer RNA enters; the first transfer RNA, bereft of its amino acid, is ejected; and the machine grinds on. Thus, a polypeptide chain is being synthesized, amino acid by amino acid, and is being extruded from the ribosome as the messenger RNA "tape" is being read, triplet by triplet.

5. The ejected, spent transfer RNAs go back into the cell at large, where each encounters another amino acid of the appropriate sort, attaches to it, and thereby becomes recharged.

As soon as the messenger RNA molecule has begun to pass through the ribosome, its leading end can enter another ribosome to begin the synthesis of a second polypeptide chain while the first one is still being produced. A third and a fourth ribosome can be entered by the RNA molecule until its entire length is dotted with ribosomes, each turning out a chain in sequence, just out of phase with the preceding ribosome on the messenger RNA molecule. Messenger RNA molecules with their attached ribosomes are shown on the facing page.

Some proteins—such as amylase (the enzyme in saliva that converts starch into sugar), or the constituent proteins of hemoglobin, or those of hair—must be produced constantly in large volume by a relatively small number of cells, each containing only one gene for each protein. This is possible because one of the main characteristics of the transcription and translation machinery is its multiplicative effect: A single master DNA sequence, the gene, will be copied by many messenger RNA molecules, and each of these in turn will be read by many ribosomes at once, each one slightly out of phase with the next. Thus, the quantity and the rate of production of a given kind of polypeptide can be very great, even though the cell contains only a single gene for that polypeptide.

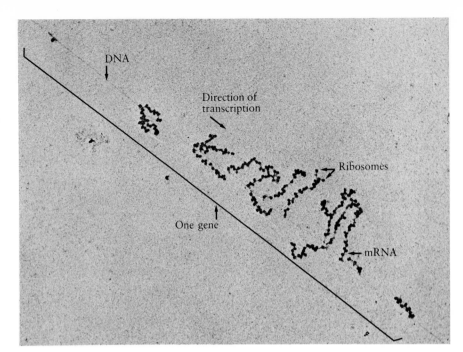

Electron microscopic image of a bacterial gene being simultaneously transcribed into messenger RNA and translated. The polypeptides and tRNA are not visible.

The Origin of Genetic Variation

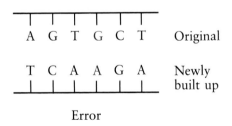

Error

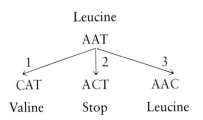

Despite the exactness with which complementary bases are matched at the time of DNA replication, errors do occur. No chemical process is absolutely repeatable. For example, sometimes opposite a G in a replicating strand of DNA, an A will be inserted in the new half-molecule instead of the C that should have been put in place there, as shown at the left. The result is a mismatch across the double helix and a strain on the molecule caused by the pairing of two large bases that do not fit into the available space. Such mismatches are unstable. The cell contains enzymatic repair machinery that detects such mismatches and repairs them by cutting out one of the offending nucleotides and replacing it with one containing the appropriate complimentary base. It is a matter of chance, however, which of the two strands is repaired. If it is the newly synthesized strand, the DNA double helix will go back to its original, correct state. But if it is the original strand that is "corrected" to match the erroneously inserted nucleotide, a completely new pair of bases will result. In our example, instead of the G-C pair that should have been at position 4, there will be a T-A pair. The DNA will be permanently changed, and this may be passed on to future generations. A mutation has occurred. Replacement of one pair by another is not all that can happen: One or more nucleotides can be lost, thus shortening the molecule, or conversely, additional nucleotides may be added.

The effect that replacing, losing, or adding nucleotide pairs may have on determining the kind of protein that is produced varies according to the particular change that has occurred in the genetic code. To illustrate the possible results, let us consider the triplet AAT, which codes for the amino acid leucine. Three possible substitutions that might occur are shown at the left. If the first occurs, the new triplet that results, CAT, codes for a different amino acid, valine. If the second occurs, protein synthesis will stop at that point because the triplet ACT is a signal to the ribosome machinery that the end of the message has been reached. As a result, an incomplete polypeptide will be produced whose length depends upon how near the front end of the gene the change occurred. If the third change occurs, nothing happens to the protein because AAC is one of the alternatives to AAT in coding for leucine.

Loss or addition of nucleotides is usually catastrophic for the protein. The translation mechanism reads the bases in the nucleotides three at a time in order from one end of the message. If a single extra nucleotide is inserted or lost, then, from that point on, the reading frame of the ribosome will be shifted *one* base to the right or the left. All the subsequent triplets of bases will be misread, and all the subsequent amino acids will be wrong, until a stop code is encountered and the entire mistaken process is terminated. It is as if all the spaces between the words on this page were shifted one place to the right producing "allt hes pacesb etweent hew ordso . . ." and so on. Obviously, the same kind of result would be produced if two nucleotides were inserted or lost. Only if three (or some multiple of three) nucleotides were inserted or lost would the rest of the code be read properly. In such a case, the loss or addition of three nucleotides would result in the dropping or adding of one amino acid in the protein chain, but the rest of the chain would be correct.

Although such events have never been observed in nature, we assume that all the genetic polymorphisms we have reviewed have arisen by nucleotide substitutions, deletions, and additions in the evolution of the human species and its ancestors. Perhaps the first members of our species were all homozygous PGM^1PGM^1 for the phosphoglucomutase gene. In Asia Minor, sometime during the last hundred thousand years—probably near the beginning, before the origin of the major human races—mutations occurred that produced the PGM^2 allele. For reasons we know nothing of, that mutation spread through the population, reaching a frequency of about 25%, its frequency still among Europeans, Africans, and Asians. Ultimately, all genetic diversity must have its origin in errors of DNA replication. Because the evolution of species depends on the existence of genetic diversity, we can say confidently that, if DNA replication were always perfect, we would not exist.

The Molecular Basis of Quantitative Variation

Clearly, it cannot be the case that all genetic variation is the result of amino acid substitutions that change the qualitative nature of our proteins. Short people, tall people, and people of intermediate height, or people with skin colors varying from dark black to the palest pink, cannot each have a different amino acid sequence for some protein. There must be heritable variation in the *amounts* of gene products as well as in the kinds of them. Moreover, developmental changes in individual people show that proteins make their appearance at specific moments and not at others. Fetal humans have fetal hemoglobin made up of α chains and γ chains, the products of the gene for α hemoglobin and the gene for γ hemoglobin. Slowly, during infancy, the γ chains stop being produced and are replaced by β chains, the product of a different gene. Somehow, the amount of product that is produced under the direction of each of the genes is scheduled and regulated. Yet quantitative variation in the production of gene products is not an internal matter alone. Height is a function both of the genes one inherits and of one's early nutritional history. Skin color depends both on genes and on exposure to the sun. Between gene and organism falls the environment.

There are two levels at which quantitative variation can occur. One is at some distance from the gene: Once enzymes and other proteins have been constructed, they enter into complex chains of reactions that are controlled by yet other reactions and by environmental circumstances. The rate at which enzymatic processes proceed depends upon the temperature and the amount of energy available. Chemical reactions—and enzymatic reactions in organisms are no exception—are concentration dependent, speeding up and slowing down as the concentrations of the reactants change. That is, genes quite aside, an organism is a complexly interconnected network of chemical reactions in intimate physical relation to the external environment.

The other level at which quantitative variation can occur is in the activity of the genes themselves. Signals from the environment penetrate to the genes and influence the rate of protein synthesis. When what Vermonters call a "flatlander" spends several weeks at high elevations, a large increase in the rate of hemoglobin production takes place to compensate for the low oxygen tension. Somehow the rate of transcription and translation of the DNA message is increased.

It is not just signals from the environment that are felt by a gene's synthetic activity but signals from other genes as well. In Mediterranean countries, there is a widespread genetic disease—Cooley's anemia, or β-thalassemia—which is marked by fragility of the red blood cells, a decrease in their size, an increase in their number, anemia, and a variety of serious related pathologies. Thalassemics die young. Despite that, the frequency of the gene that causes β-thalassemia is quite high (about 10%) in affected populations, and it must, therefore, be classed as a major polymorphism. The genetics of β-thalassemia is simple. There is a single gene segregating with a normal and an abnormal allele. Homozygotes for the abnormal allele suffer the disease, while heterozygotes have only a milder form of the abnormality. Examination of the hemoglobin of β-thalassemics shows that the amino acid sequences of their α, β, and γ chains are perfectly normal. However, these normal chains occur in abnormal amounts. There are very few β chains produced, and so there is a large excess of

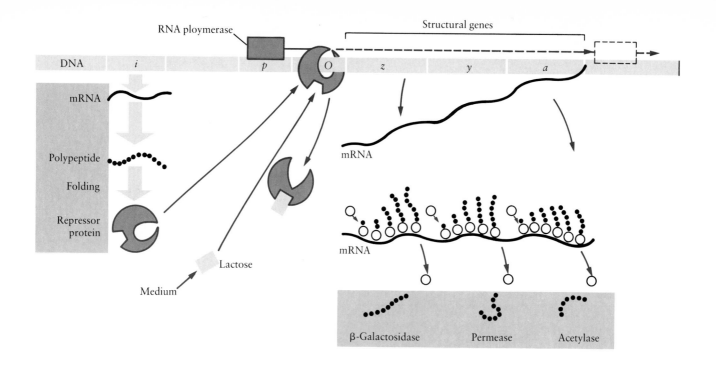

Regulation of the lactose operon. The *i* gene (which is not considered a part of the operon and is somewhat separated from it) continually makes repressor. The repressor binds to the *O* (operator) region, which prevents the RNA polymerase bound to *p* from transcribing the adjacent structural genes. When lactose is present, it binds to the repressor and changes its shape so that the repressor no longer binds to *O*. The RNA polymerase is then able to transcribe the *z, y,* and *a* structural genes, and the three enzymes are produced.

α chains. As fetal hemoglobin production is shut down in early childhood, there are not enough β chains to take the place of the vanishing γ chains, and so normal adult hemoglobin cannot be produced in sufficient quantities. The β-thalassemia mutation is an example of a regulatory change that acts to control the rate of synthesis of the product of a gene.

Mechanisms of gene regulation vary, but a common one is the mechanism that is typified by the regulation of the production of the enzyme β-galactosidase in bacteria. This mechanism is shown above. The entire structure of gene function is called an *operon*. On the right is the sequence of DNA that codes for the β-galactosidase molecule itself and two other enzymes, used in lactose metabolism. Just to the left of this structural gene is a special stretch of DNA—the operator, or *O*, region—and next to it is another special sequence—the *promoter*, or *p*, region. Yet further away (in some similar cases, in a completely different part of the chromosome) is another structural gene, the *i* gene, that codes for a special protein, the repressor protein. The mechanism is simple: Making a messenger RNA copy of the gene's DNA requires the presence of an enzyme, RNA polymerase. This enzyme attaches to the DNA strand at the promoter region and then moves along it, putting together the strand of messenger RNA. In order to do so, it must pass across the *O* region. The *O* region has a special affinity for the repressor protein, which can sit there blocking the movement of the polymerase and thereby preventing gene transcription. Thus, gene transcription is simply "turned off" by the repressor protein. The repressor protein, however, has a flexible structure. If that structure is distorted, the protein will not fit into the O region, and the RNA polymerase will not be blocked. Therefore, the gene will be "turned on." In general, the repressor molecule becomes distorted by being attached to a small molecule that is part of the biochemical pathway with which the gene is concerned. In the figure on the facing page, the small molecule is lactose—the same kind of molecule that is acted upon by the enzyme β-galactosidase, which is being produced by the gene. The functional chain works as follows: When lactose molecules are present in the cellular environment, they will enter the cells. They will combine with the repressor molecules, thereby changing their shape and preventing them from attaching to the O region. The β-galactosidase operon is now turned on, and the RNA polymerase is now free to move across it, making messenger RNA. The messenger RNA is translated by the ribosomes into β-galactosidase and the two other enzymes of the operon, which then break down the lactose to provide energy for the cell. Once the lactose is all used up, the repressor molecules are no longer distorted; they are able to attach themselves to the O region, thereby turning the gene off. In this way, the synthesis of proteins can be turned on and off by signals from the environment.

It has been suggested that hemoglobin production is increased at high altitudes by just such a control mechanism. The decrease in oxygen in the blood presumably distorts a repressor molecule, causing it to fall off an O region, thereby turning on the genes that code for hemoglobin. Genetic variations in the quantity of hemoglobin might also be the result of alterations in p, O, or i regions. A promoter region may be defective, for example, or a repressor may be so tightly bound to an operator region that it is virtually never removed.

The particular details of gene regulation are different for different genes and for different organisms. An alteration in the molecular environment of a cell may simply alter the stability of the messenger RNA itself, without influencing its rate of production. The central point is that the multiplication of the message in the master DNA sequence through the production of messenger RNA, and the further multiplication of that message through multiple simultaneous translations of each messenger RNA molecule by the ribosomes, makes possible the control of the rate of production of protein molecules by feedback of information from the environment. Thus, the activity of genes is both protected from the environment and sensitive to it. It is protected from the environment in that the qualitative nature of the gene product, the amino acid sequence, is built into the DNA sequence itself. It is sensitive to the environment in that the gene's production of protein can be controlled by specific signals from the environment. The organism is neither the inevitable unfolding of an internal program nor the unconstrained mirror of the environment. It is the unique product of the interpenetration of internal and external factors.

Continuous Variation

6

The variation we see around us when we walk out among people in the street is very different from the variation in blood groups or enzyme proteins described in Chapter 3. There are no neat classes (like A, B, AB, and O) into which people can be distributed on the basis of their appearance. People come in a continuous array of heights, weights, and colors, and they show extensive, subtle variation in their noses, ears, hairlines, and postures. Even those differences that are normally spoken of as qualitative—like "black" and "white"—show continuous variation when given a second glance. Eyes do not come in only two colors, blue and brown. Take any ten Europeans at random and they will almost certainly be distinguishable one from another by their eye colors, just as ten randomly chosen "blacks" will vary in skin color enough to be told apart on that basis alone. The problem is to analyze the biological, social, and environmental factors whose joint actions give rise to this continuous variation. To do so requires, at the beginning, basic methods for describing and summarizing the variation.

Basic Statistical Notions

If the heights of a sample of adults in Boston are measured, say to the nearest 5 centimeters, a few of them will be quite short, say between 156 and 160 centimeters (roughly between 5 feet 1 inch and 5 feet 3 inches) and a few will be quite tall, between 186 and 190 centimeters (6 feet 1 inch to 6 feet 3 inches), but most will be a middling height, in the neighborhood of 171 to 175 centimeters (5 feet 7 inches to 5 feet 9 inches). The number of people in each height class can be recorded in the form of a bar graph, like the first figure on the next page, which is a representation of the *frequency distribution* of heights in the population sampled. In general, such a frequency distribution is a listing of the numbers (or proportion) of people in a population that fall into each measured class. The exact shape of the distribution obviously depends upon how fine the measurements are. If the same population were measured to the nearest centimeter, the distribution would appear as in the second figure. Because there are five times as many classes as before, the absolute number of measurements in each class must be smaller than that in the first figure, but we can make up for that simply by measuring five times as many Bostonians. Conceivably, we could continue to refine our measurement to tenths and then to hundredths of a centimeter, each time increasing the number of people measured, so that we finally obtain a virtually continuous curve like the one in the third figure. This last curve is a close approximation to the continuous frequency distribution of height in a very, very large population.

Frequency distributions vary in shape, and it would be helpful to have some simple way of characterizing them that would enable us to compare them without giving the distributions in detail. Two obviously useful characteristics of a distribution are, first, its general location along the axis of measurement—that

Frequency distributions of heights of male adults when measured to the nearest 5 centimeters (top) and when measured to the nearest centimeter (middle). The bottom graph is a continuous approximation of the bar graph above it.

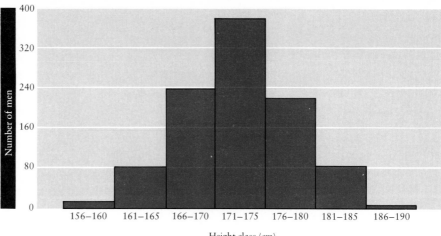

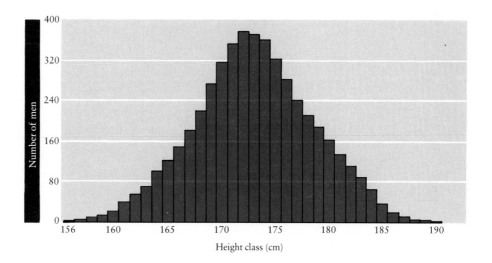

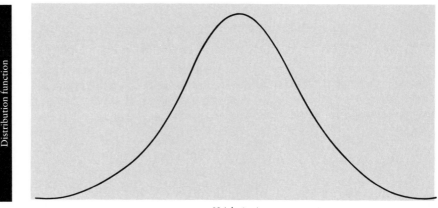

A comparison of the frequency distributions of heights for Pygmies, Bostonians, and Dinkas.

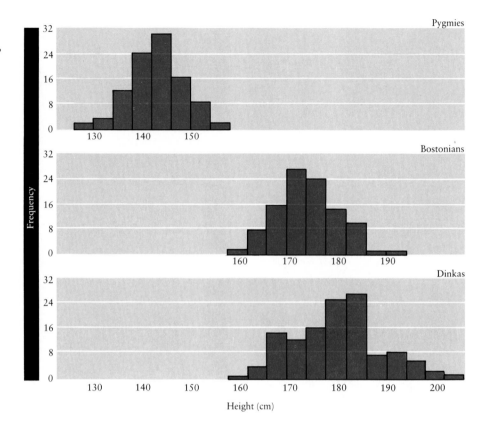

is, its central tendency—and, second, how spread out around the central value it is. Whereas the heights of Bostonians center on 170–175 centimeters, the heights of Pygmies in the Congo have a similar distribution but shifted downward to center on 145–150 centimeters (see the illustration above). In contrast, the very tall Dinkas of Africa center on 180–185 centimeters. The height distributions of Bostonians and Pygmies are very similar in spread around their central values, the tallest and shortest within each group differing by about 35 centimeters. The Dinkas, however, are not only taller but much more variable with a range of about 50 centimeters from tallest to shortest. The height distribution of the Dinkas is generally flatter and less peaked.

There are several ways of characterizing central tendency that give slightly different information about the distribution: In popular usage, they are often confused, but they really can be quite different numbers. The *mode* of a distribution is, as its name suggests, the value that occurs most frequently. For the heights of Bostonians, the mode (when values are measured to the nearest centi-

meter) is 172 centimeters. The *median* of a distribution is the middle value, such that half the people measured are above it and half are below it. The median height of the Dinkas is 178 centimeters. The *mean*, which is the most commonly used measure of central tendency, is simply the arithmetic average of the measurements, calculated by adding up all the measurements and dividing the sum by the number of people measured. The mean height of the Dinkas is 180.3 centimeters.

For a roughly symmetrical distribution like the one for Bostonians, the three measures will be very close numerically, but they will be quite different if the distribution is very asymmetrical with a long tail on one side. For example, the distribution of income in the United States is quite asymmetrical, with a few people making a very high income, while most people make much less. The modal declared income in 1977 was $18,000 but the median was only $10,000 (that is, half of all incomes were less than $10,000) and the mean was $15,000. The mean is so much higher than the median because a few people with huge incomes raise the mean considerably while having virtually no effect on the median. The usual measure of the spread of values around the center of the distribution is the *variance*, defined as the arithmetic average of the squared differences between the observations and the mean. If there are many people who deviate a great deal from the mean in either direction, the variance will be large, whereas, if all values cluster strongly about the mean, the average of all the squared differences will be small. The Pygmies and the Bostonians have about the same variance in height, 31.8 square centimeters and 37.9 square centimeters, respectively. The variance of the Dinkas' flatter, more spread out distribution is twice as large, 89.5 square centimeters. One inconvenience of the variance is that, by its method of calculation, it comes out in squared units (square centimeters, in this example). It is more appealing to have a measure of variation that is in the same units as the measurements themselves. For this purpose, we generally use the square root of the variance, called the *standard deviation*.

Both the mean and the variance of distributions are relevant for describing variation, but at different levels of comparison. The simplest description of the diversity of height among the three groups considered here—Pygmies, Bostonians, and Dinkas—is the statement of the mean heights of the three groups. However, the description of the diversity *within* any one group is either the variance or the standard deviation of that group. Suppose that, in addition to Bostonians, Pygmies, and Dinkas, we had measured 97 other local populations. Each would have a mean height. The description of variation among the 100 groups could, of course, be a list of the 100 means, just as the description of variation within a group could be the list of individual heights. Alternatively, we could treat each mean as if it were an individual value and then calculate the variance among those means. The resulting single number, the variance of the means, would express how different the populations are one from another.

The variance of the individual measurements within each population and the variance of the means measure two very different things and answer two very different questions. The variance of the means tells how different populations are from one another. The variance of individual measurements tells how different individuals are from one another *within* each population. There may be lots of variation within a population but no difference between groups. There is essentially no difference in height between Spaniards and Portuguese, but there is a good deal of variation from person to person within each population.

The Sources of Population Variation

There are three sources of phenotypic variation among individual people. First, they differ genetically one from another. Second, people have different environmental and social histories. Third, there are accidents of development quite aside from external environmental fluctuation.

It is important to realize that variation in any or all of these factors does not necessarily mean that there will be manifest variation in the phenotypes of people. Every person differs from every other person in *some* genes but not necessarily in those genes whose variation is relevant to a particular trait, such as height. Moreover, people may differ in genes that are relevant to the development of a particular trait, but, in the environment in which those people actually live, no variation in that trait can be seen among them. Let's assume, for example, that there is a gene whose function is to regulate fat deposition. Different forms of such a gene would make for different rates of deposition. If people were to have a great deal to eat or were to eat very starchy diets as young children, the differences in their genotypes would show up as differences in body fat. If, however, everyone were getting barely enough calories to meet daily energy needs, everyone, irrespective of genotype, would be thin. Genetic differences are turned into phenotypic differences in some environments but not in all of them.

Variation in the environment and developmental errors may also fail to be translated into manifest variation in a trait, even when the environment and the developmental process are directly relevant to the trait. All organisms, and human beings more than most, possess the property of *homeostasis*—the buffering out of disturbances to their physiology and development so that they maintain a constant function or form despite environmental variation. Over a broad range of altitudes, people will maintain a constant supply of oxygen to their tissues by means of a complex series of homeostatic devices, including modifying the heart rate and the breathing rate and, eventually, changing the number of red blood cells. Thus, the variation in a very important environmental factor is, in effect, neutralized by the organism. Yet, this neutralizing ability depends on the genotype. Some genotypes can do it and some cannot. People with normal hemoglobin A can withstand high altitudes, but heterozygotes for the sickling gene (genotype $Hb^A Hb^S$) will suffer at high altitudes because of reduced oxygen tension. Their mixed hemoglobin will crystallize when the oxygen tension gets low enough, their red blood cells will sickle, and they will suffer an attack of anemia.

In general, then, the sources of variation between people interact. Variations in genotype may or may not be manifest as variations in physiology, development, and behavior, depending upon whether the environment translates those genetic differences into manifest variation. Conversely, variation in environment may appear as variation among people if they have one kind of genotype, but not if they have another. Genotype and environment are inextricably bound up together in producing the whole organism and its variation.

Genetic Variance and Environmental Variance

Consider the frequency distribution of heights in a population. If the population has different genotypes influencing height, then that distribution is really a composite of a large number of distributions mixed together in the population as a whole: Each of those subdistributions describes one of the genotypes in the population. Some genotypes have small mean heights and others have large mean heights. Also, each genotype has a distribution of heights around its

Distribution of red-cell phosphatase activities in the general population (in color) and in the separate genotypes. The curves are constructed from data on the frequencies of the different genotypes in the English population.

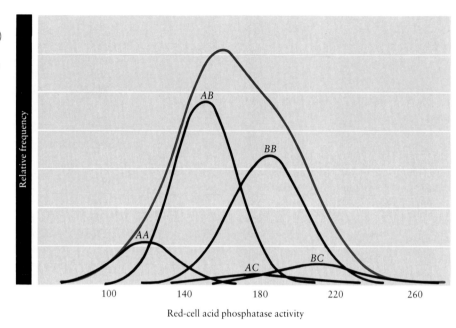

Red-cell acid phosphatase activity

mean, because different people of that genotype have developed in different environments. If one genotype is very sensitive to the environment, there will be a large variance around its mean. If another is relatively insensitive to the environment, all people of that genotype will have heights close to their genotypic mean. The shape of the distribution for the total population is simply the consequence of adding together the individual distributions, each with its own mean and variance.

Looked at in this way, the spread of the distribution for the total population is derived from two sources. First, the means of the underlying genotypic distributions are different. Therefore, the population will have a variance just because there is variance among the genotypic means. This variance among genotypic means is the *genetic variance* in a population. But the variation among genetic means does not account for all the population variation: There is also variation within each genotype, as shown by the individual genotypic subdistributions. This variance *within* each genotypic class is called the *environmental variance*, because it would disappear if the environment were identical for all individuals.

A concrete example of how a frequency distribution for a total population is composed of the underlying distributions of different genotypes is shown in the graph above. The curve in color is the distribution of red-cell acid phosphatase activity in the English population. The gene coding for the enzyme red-cell acid

Red-blood-cell activity of different genotypes of red-cell acid phosphatase in the English population

Genotype	Mean Activity	Variance of Activity	Frequency in Population
AA	122.4	282.4	.13
AB	153.9	229.3	.43
BB	188.3	380.3	.36
AC	183.8	392.0	.03
BC	212.3	533.6	.05
CC	~240	—	.002
Grand average	166.0	310.7	
Total distri- bution	166.0	607.8	

Source: H. Harris, *The Principles of Human Biochemical Genetics*, 3rd ed. (North-Holland, 1980).

phosphatase is polymorphic with three alleles, *A*, *B*, and *C*. There are thus three kinds of homozygotes, *AA*, *BB*, and *CC*, and three kinds of heterozygotes, *AB*, *AC*, and *BC*. A person's genotype can be determined by electrophoresis of a blood sample. Each genotype also makes for a different degree of enzymatic activity. People of genotype *AA* have the least enzymatic activity, those of genotype *CC* have the most, and those of genotype *BB* are in between. Further, each heterozygote has an activity halfway between the relevant homozygotes. Thus, the mean rates of enzymatic activity in the blood of people of genotypes *AA* and *BB* per unit time are 122.4 and 188.3 units, whereas that of people of genotype *AB* is 153.9 units, almost precisely halfway between. The mean activities are shown in the table at the left. Different people of the same genotype, however, do not all have the same activity. There is a distribution for each genotype, the variance for which is given in the next to the last column of the table. Each of the numbers in this column is the environmental variance for its respective genotype, and it should be noted that the numbers differ from genotype to genotype, with *BC* having nearly twice the variance of *AA*. The last column in the table shows the frequency of the genotypes in the English population. Homozygotes *CC* are so rare that a reliable mean and variance for their enzymatic activity could not be determined.

If you now look back at the figure on the facing page, you will see how the total distribution is composed of the sum of the underlying distributions. Each genotypic distribution is drawn to a scale proportional to its frequency in the population. Thus, the largest distribution is *AB*, because it is the most frequent, while the smallest is *AC*. The total distribution is constructed by adding up the underlying distributions.

The total variation in the population comes from two sources. One is the variation among the means of the distributions of genotypes *AA*, *AB*, *BB*, *AC*, and *BC*. The other is the variation within each of those genotypes. Thus, people with less than 100 units of activity are almost certainly of genotype *AA*, but at the low end of the activity scale for that genotype. On the other hand, people with 170 units of activity may belong to any one of four genotypes: *AB*, *BB*, *AC*, *BC*. People of genotype *BB* have a fairly good chance of having a rate of activity anywhere between 150 and 230 units, about half the range of the entire population. Numerically, the variance of the total distribution turns out to be 507.8, but the average of the environmental variances within each genotype is only 310.7 Thus, the variance among the means of the genotype, the genetic variance, 607.8 − 310.7 = 297.1, accounts for almost half of all the variation in the total population.

The case of red-cell acid phosphatase illustrates a number of basic phenomena of quantitative variation. The first is that the genotype does not determine phenotype: A large range of phenotypes exists for each genotype. The second is that genotypes vary in their sensitivity to the environment, as evidenced by their different environmental variances. The third is that variation in the population as a whole will change if the frequencies of the genotypes change. Suppose that, for some reason, the *B* and *C* alleles of the gene were to disappear, leaving only *AA* homozygotes. The mean of the population would decrease from 166 to the mean of *AA* homozygotes, 122.4. But the variance would also decrease: All the genetic variance would disappear, and the remaining variance would be only the environmental variance of *AA* homozygotes. Thus, when we speak of the genetic variance or the environmental variance of a trait, we are being inexact. The variance is a characteristic not of a trait but of a trait in a particular population of genotypes in a particular set of environments. If either the frequencies of the genotypes or the range of environments is changed, the variance within the population will also change. The variability of a trait is then a consequence of the biological history of the population (as expressed in the frequencies of the different genotypes) and of the social organization of the population (which determines the kinds of environments to which the individuals are subject).

The Interpenetration of Genotype and Environment

"Words are wise men's counters, they do but reckon with them, but they are the money of fools." Hobbes's aphorism applies as well to variance. Genetic variance has been defined as the variation among genotypic means in the population, and environmental variance has been defined as the average variation within genotypes in the population. It would be a mistake to suppose, however, that "genetic" variance is sensitive only to changes in the genes while "environmental" variance is changed only by changing environments. In fact, this separation of variances does not really separate genetic and environmental causes of variation, for they are, in one sense, inseparable.

It is easy to see that the environmental variance depends on the genotype. In the figure on page 66, the environmental variance of the entire population is 310.7, which is the average of the environmental variances of all the genotypes in the population, weighted by their frequencies. Suppose that alleles *A* and *C* were to disappear from the population: All members of the population would be of genotype *BB*, and the environmental variance would be 380.3. Thus, a change in the *genotypic* composition of the population would change the population's *environmental* variance.

It is a little more complicated to show that genetic variance, in its turn, depends upon environment, but this sensitivity plays a vital role in our understanding of the implications of different kinds of variation. Consider the upper graph on the next page, which shows the speed of reaction of two genetic variants of the enzyme glucose-6-phosphate dehydrogenase (G6PD) as a function of

Illustration of how genetic variance of a trait may appear and disappear from the population as the environment changes. *Top:* Norms of reaction for two different forms of the enzyme glucose-6-phosphate dehydrogenase as a function of the concentration of the substrate on which the enzyme works. *Bottom left:* The population distribution of enzyme activity when the environmental range is given by the range "L" in the upper graph. *Bottom right:* The population distribution of enzyme activity when the environments are in the range "R."

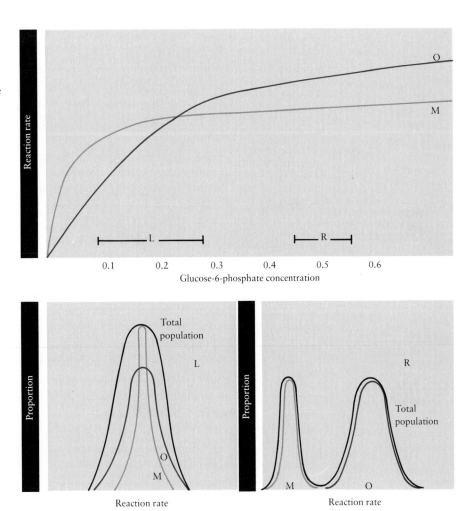

the concentration of glucose-6-phosphate. These are norms of reaction (Chapter 2): They show the phenotypic characteristic (reaction speed) of each genotype as a function of environment (concentration of reactant). Note that the curves representing the two norms of reaction cross: Variant O is more active at high concentrations, while variant M is more active at low concentrations. Moreover, the environmental sensitivities of the two genotypes are different. Over a large part of the range of environment, genotype *M* has virtually a constant reaction rate, while genotype *O* continues to increase its rate with increasing substrate concentration. Suppose now that the actual distribution of glucose-6-phosphate in cells of different people were in the range delineated by the right-hand bar, R. In this range of environment, there is a big difference between the genotypes: The reaction rate of variant O is clearly higher than that of variant M. There will not be much variation among people of genotype O and virtually none among those of genotype M because, in this range of environments, both genotypes are fairly insensitive to concentration. What would the activity distribution in the population as a whole look like? It would consist of two clear modes, one corresponding to people of genotype O and one to people of genotype M, with no overlap between them, as shown in the distribution on the lower right. There is a large amount of *genetic* variance because the means of the two genotypes are very different.

Suppose the environment were to change to the left-hand range, L. In this range, there is no difference between the genotypes on the average, although genotype O is more sensitive to environmental variation than genotype M. As a consequence, the composite distribution for the population will look like the one on the lower left. There is a single mode, and there is no genetic variance. Thus, the *genetic* variance has been changed—indeed, destroyed completely—by changing only the *environment*.

The consequence of the sensitivity of genetic variation to environment is that simple predictions that seem reasonable at first glance may well turn out to be wrong. If in some population the genetic variance for a trait is very high while the environmental variance is low, one might be tempted to say that it is useless to change the environment as a way of changing the trait. After all, the trait seems to be sensitive to differences in genotype but not to differences in environment. But, as our example has just shown, this might be quite wrong. A simple shift in the environment could change the mean of the entire population and get rid of all genetic variation. Would one be justified, seeing only a situation of this latter sort, in affirming the irrelevance of genotype and the supreme importance of environment? No, because the two genotypes really are quite different if the entire possible range of environments is taken into account.

In general, it is not possible to extrapolate from one set of environments to another using only information about sensitivity to environment within one narrow range. Traits that are totally insensitive to environmental change in one range of environments may be quite sensitive to it in another. Conversely, even if different genotypes are, on the average, very different in one environment, they may be identical in another.

Heritability

Such variable traits as red-cell acid phosphatase activity, in which it is possible to identify simple genetic differences underlying the variation, are very rare. Usually all that can be observed is the variation in the trait itself and the pattern of that variation in families. These family patterns are not simple, however.

Matings between people 170 centimeters tall and those 160 centimeters tall will produce offspring of a whole variety of heights. It is not possible, simply by looking at parents and their offspring, to deduce their genotypes. Indeed, if all we had to go on were the enzyme activity of parents and their offspring, the variation of the red-cell acid phosphatase gene could not have been reconstructed either. A glance at the figure on page 66, will show this. A mating between two people with 180 units of activity might have been a mating between an *AB* and a *BB,* or an *AC* and a *BC,* or any combination of those four genotypes, and the enzyme activity of the offspring could have fallen anywhere in the entire activity range. Simple genetic analysis, which is fairly easy for such traits as blood types, is just not possible for quantitative traits. Indeed, many quantitative traits probably have no genetic variation at all underlying them. It is hardly likely that there are genes whose allelic variation influences the ability to speak French as opposed to German, or that there are genes that are relevant in any direct way to our attitudes about the Incarnation. The most one can do with most quantitative traits is to ask whether there is any evidence at all that genes influence variation in them and, if so, how much genetic variation relevant to the trait in question exists in a given population. Specifically, we can ask the question, "What proportion of the variance of a particular trait in a given population is genetic variance?" This genetic component of the variance is the *broad heritability* of a trait in a population, usually called simply *heritability* and symbolized H^2. The heritability of red-cell acid phosphatase activity in the English population is

$$H^2 = \frac{\text{genetic variance}}{\text{total variance}} = \frac{297.1}{607.8} = 49\%.$$

The heritability of MN blood type is 100%, and the heritability of the ability to pronounce *rue* like a Frenchman is zero.

Because heritability is simply the ratio of genetic to total variance for a trait, all the pitfalls and caveats associated with genetic and environmental variance apply to H^2 as well. Most important, heritability is a characteristic not of a trait but of a trait in a particular population in a particular set of environments. The same trait may have different heritabilities in different populations in different environments. The heritability of a particular trait in a particular species might be zero because no genes in that species that are relevant to the determination of the trait are variable. But the heritability of a trait might also be zero in a particular population because all of the individuals in that population happen to be homozygous for one allele of the relevant gene while other populations might have variant alleles. Or, it might be that, in the environment of one population, the various genotypes have the same mean expression, whereas in a different environment their average expressions would differ. The lack of heritability does not reveal the cause of that lack. Conversely, a high heritability of a trait in one population at one time tells us nothing about its heritability at other times in other populations.

A second important fact about heritability—one that is widely misunderstood—is that knowing the heritability of a trait cannot help one to decide how to change it. It is simply not true that, if a trait has a heritability of, say, 90%, it is useless to change the environment because somehow "genes determine the trait." Our discussion of genetic variance showed that it is not possible to extrapolate from the amount of genetic variance to the consequence of a change in the environment. *All that the heritability of a trait tells us is how much genetic variation exists for that trait at a particular time in a particular population.* A measure of heritability contains no implicit prescription for change.

These cautions about heritability are not academic quibbles. They lie at the heart of important social issues. A great deal of effort has been devoted to trying to determine "the" heritability of such quantitative human traits as IQ performance, wealth, schizophrenia, mental retardation, blood pressure, and so on, in the belief that a knowledge of the heritability of these traits will somehow prescribe social action. A recent U.S. court decision held that claimed cures for baldness are necessarily fraudulent because baldness is "inherited". But there are examples whose import is much more serious: It has been asserted that compensatory education will necessarily fail because IQ is said to have a high heritability. Similarly, it has been asserted that psychiatric treatment of schizophrenia either is useless or must necessarily depend upon drugs and other physical manipulations because schizophrenia is "heritable". In each case, the same fallacy is operating: "Heritable" is taken to mean "insensitive to environmental change." In addition, in the case of mental disease, the implication is that, because it is heritable, there must be a molecular defect and that, therefore, only a molecular manipulation can treat it. In part, the problem is that a word like "heritability" carries with it an everyday meaning that has been confounded with its technical meaning. Counters have become money. But the problem is deeper. It goes back to the false dichotomy between nature and nurture, to the belief that gene and environment are separate and separable determinants of organisms rather than interacting and inseparable shapers of development.

The Estimation of Heritability

All studies of inheritance, whether of quantitative traits or of simple qualitative traits, are studies of the resemblances between relatives. If genetic differences are implicated in differences between phenotypes, we expect relatives to look more alike than unrelated people, since relatives are likely to have some of the same genes, inherited from common ancestors. Because parents pass their genes on to their children, we expect those children to look more like each other and more like their common parents than they look like the children next door. The basic methodology of genetic investigation is always such family comparisons. The trouble is that members of the same family share more than some of the same genes; they share some of the same environments as well. To get at the role of genes in shaping family resemblances, it is necessary, somehow, to cancel out the effects of environmental similarities. In experimental organisms, that is no problem: Several offspring of the same pair of parents are raised in controlled, separated environments together with offspring of other parents. Cows and mice can be removed from their mothers at birth and nursed mechanically or given to foster mothers. Seeds of plants can be grown in rigidly controlled environments. But the nuclear or extended family and social class are realities of human life that cannot be so easily cancelled out or manipulated by experimenters. As a result, there is a complete confounding of the effects of genetic similarity and those of environmental similarity on the resemblances among parents and children, uncles and nieces, brothers and sisters.

It is essential to distinguish between family resemblance and genetic resemblance. Many traits are *familial* but not *heritable:* For example, in the United States, the greatest similarity between parents and offspring is in two social traits, religious sect and political party. Yet no serious person would suggest that the very high family resemblance for these traits is a result of genetic determination.

Familiality is often confused with heritability, when it is supposed that the resemblance of parents and children is a demonstration of the power of heredity. Familial similarity is the *observation*. It should not be confused with the *explanation,* which may involve genetic and environmental commonalities. The cases in which people confuse the two can be deeply revealing of prior social assumptions. Although no one takes a similarity in the political affiliation of parents and children to be evidence of a gene for political party, it is widely believed that a similarity between parents and offspring in IQ scores is *prima facie* evidence that genes influence—even determine—intelligence. Alcoholism is commonly thought of as inherited because it often happens that a father and his son are both known to drink excessively, but we never hear it said that Presbyterianism is in the genes. Yet the evidence is really the same. Assertions that alcoholism might somehow have a simple biochemical basis confuse physiological sensitivity to a given amount of alcohol, which does indeed appear to be heritable, with a complex social behavior—drinking compulsively—for which there is no evidence of heritability.

The only solution to the dilemma of environmental similarity is, in principle, to conduct adoption studies. Children separated from their parents and their

siblings at birth can be compared with their foster relatives and their biological relatives. If a trait is completely heritable, we would expect adopted children to resemble their biological relatives closely with respect to that trait, while they ought to be no more similar to their foster parents than to randomly chosen people. On the other hand, we would predict that adopted children would resemble their foster parents, rather than their biological mother and father, in such traits as religious affiliation. Thus, adoption studies lie at the heart of human quantitative genetics. At first glance, the most seductive of these studies are those of identical twins raised apart. Because identical twins result from the separation of a single fertilized egg into two complete organisms, they are genetically identical. Therefore, if they are similar even when raised apart, their similarities must be the result of common genes; their differences will reveal the effects of environmental variation. The trouble is that there are practically no cases of identical twins who have been raised in truly different environments. In the next chapter, we will review the evidence about these romantic lives. But it is sufficient, for the moment, to observe that, when twins are separated at birth in real circumstances, they usually turn out to be given to very similar families.

A Technical Note on the Measurement of Similarity

The standard measure of similarity between two series of values is the *correlation coefficient*. It is calculated from the deviations of each variable from its own mean. Consider the measurements in the table below. Person C is 1 centimeter below the mean height and 5 millimeters below the mean foot length.

Person	Height (cm)	Deviation from Mean	Foot Length (mm)	Deviation from Mean
A	170	−4	270	−20
B	171	−3	275	−15
C	173	−1	285	− 5
D	175	+1	295	+ 5
E	177	+3	305	+15
F	178	+4	310	+20
Mean	174		290	

Person B is 3 centimeters below the mean height and 15 millimeters below the mean foot length. In fact, each person's deviation from the mean height is exactly proportional to that person's deviation from mean foot length. In this artificial example, height and foot length show perfect positive correlation: Their correlation coefficient is +1.0. If one of the columns were turned upside down so that positive deviations in height matched negative deviations in foot length, the variables would show perfect negative correlation, with a correlation coefficient of −1.0. If the columns were scrambled so that the variables bore no regular relation to each other, the correlation would be nearly 0. For example, scrambling the foot-length column as shown at the left gives the smallest positive correlation possible with these particular numbers: +.057.

Correlation is a measure of how similar the *deviations* in one variable are to the *deviations* in another. It is not a measure of actual identity. Notice that, in the first example, foot length and height are perfectly correlated, yet the numbers are not the same in the two columns. Two variables can be perfectly correlated even when every value of one is larger than every value of the other. This fact is relevant to an understanding of the meaning of correlations in adoption studies: Even if the traits of a group of children are perfectly correlated with those of their biological parents and completely uncorrelated with those of their adoptive parents, it is entirely possible that those children, *as a group*, resemble their adoptive parents very significantly. Let us consider the hypothetical (but nevertheless realistic) set of observations in the table on the next page: the heights of children from impoverished Guatemalan peasant families, the heights of their biological parents, and the heights of their adoptive, middle-class North American foster parents. There is a perfect positive correlation between the

Person	Height (cm)	Foot Length (mm)
A	170	310
B	171	275
C	173	270
D	175	295
E	177	305
F	178	285

Height (cm)		
Children at Adulthood	Biological Parents	Adoptive Parents
170	160	178
171	161	171
173	163	170
175	165	175
177	167	177
178	168	173

heights of the children and those of their biological parents, but there is no correlation between the heights of the children and those of their adoptive parents. Thus, we are entitled to conclude that height is highly heritable. However, all of the children are a full 10 centimeters taller than their biological parents. The explanation is that the superior nutrition the children received in their adoptive homes resulted in considerable growth, but equally in all the adopted homes. As a consequence, they showed no person-by-person similarities to their adoptive parents, but the general amelioration of their environment made them as tall, on the average, as their foster parents and much taller than their own malnourished biological mothers and fathers.

The phenomenon illustrated by the growth of these hypothetical children is directly related to the issue of heritability and environmental similarity: Height, in this example, is totally heritable. The heights of the children are perfectly correlated with those of their biological parents. Yet that perfect heritability does not contradict the possibility of making the children grow taller by giving them better nutrition.

The Heritability of Some Human Traits

It should come as no surprise that very little is known about the genetic variation underlying those human traits that show continuous variation. The kinds of adoption studies that would be needed to distinguish mere familiality from real heritability are not common, partly because the data are expensive to acquire but also because adoptions are usually not truly random with respect to family environment. There is a lot of folklore about the heritability of various human traits but not a great deal of good evidence.

An example of the difference between everyday wisdom and science is the case of musical ability. It is widely believed that musical ability is "inherited," yet, when the evidence is examined, it evaporates. First there is the problem of defining "musical ability." Is it the ability to compose, or to perform on an instrument, or simply to carry a tune, or just to be able to distinguish ascending from descending sequences of notes? The evidence offered in support of the notion that outstanding musical ability is inherited is usually the datum that composers run in families. After all, look at the seven generations of musical Bachs, of whom two were truly outstanding, the Mozarts (father and son), the Scarlattis, or the Haydn brothers. But anecdotes are not evidence. For every Bach family, there are scores of outstanding musicians who were the first and last of their line. Mendelssohn's father was a banker, Chopin's a bookkeeper, Schubert's a schoolmaster, Haydn's a wheelwright. Much less is known about their mothers, but none were known composers or performers. It is not at all clear that the number of musical families relative to the number of musical singletons is greater than might occur by chance, because no one has ever com-

Johann Sebastian Bach (upper left) and the three most famous of his musical children (clockwise), Johann Christian, Wilhelm Friedeman, and Karl Phillip Emmanuel, the last of whom was considered in his day to be *the* Bach.

piled the statistics of familiality of musical performance and composition. In reading about the lives of composers, one certainly gets the impression that a large number were the children of minor performers—mediocre court tenors like Beethoven's father or free-lance double-bass players like the elder Brahms. Even granting familiality, it is impossible to say whether the correlation between parent and offspring is genetic in any part or is entirely environmental. The familiality of musical performance certainly seems to have been greater in the eighteenth and early nineteenth centuries than it is now, which is what one would expect from a time in which music was a trade like tailoring, tinkering, or baking, passed on from father to son as a way of making a living. Unfortunately for science, the prolific Johann Sebastian Bach did not have the foresight to have ten of his twenty children fostered from birth by pastry cooks and peasant families so that we might estimate the heritability of musical genius.

Heritabilities of human traits

Trait	H^2
Height	.94
Weight	.42
Arm length	.87
Foot length	.81
Hip circumference	.66
Cephalic index	.70
(head breadth/head length)	
Masculinity-femininity	.85
IQ	.53
Extroversion	.50
Neuroticism	.30

Source: L. L. Cavalli-Sforza and W. Bodmer, *The Genetics of Human Populations* (W. H. Freeman and Company, 1971).

The table at the left shows estimates of heritability, H^2, for a number of physical and personality traits of white women in the United States. These estimates are from a number of different studies. The physical measurements are clearly defined. The "personality" traits, however, refer to scores on various standardized tests or the outcomes of structured interviews. The reader should not take too seriously the names given these tests. Whether "masculinity," "neuroticism," or "intelligence" has been assessed is a matter open to debate, inasmuch as there is no agreed upon definition or measure of such constructs.

The values given in the table have not been estimated from adoption studies. Rather, they depend upon an alternative approach, using twins, that is meant to solve the problem of environmental correlation. Identical, or *monozygotic,* twins are genetically identical since they come from a single fertilized egg. Fraternal, or *dizygotic,* twins are simply siblings who happened to have been conceived simultaneously from the fertilization of two eggs. They are no more alike genetically than any pair of siblings: They share 50% of their genes. If we can assume that the environmental correlation between monozygotic twins is the same as that between dizygotic twins, since a pair of twins of either sort is brought up as a pair in the family, then we can use comparisons between the two kinds of twin pairs to estimate heritability. Identical twins will be identical for a trait with 100% heritability, while fraternal twins will not. In general, symbolizing the correlation between monozygotic twins as r_M and between dizygotic twins as r_D, an estimate of heritability is:

$$H^2 = \frac{r_M - r_D}{1 - r_D}$$

At all ages, monozygotic twins have their resemblances reinforced by being dressed alike and by following the same life activities.

It was this formula that was used to obtain the values in the table on the facing page. It is exceedingly unlikely, however, that the assumption of the method is correct. Dizygotic twins, even when they are of the same sex (as they are in all such calculations), are not treated like monozygotic twins. The very similarity in physical characters between identical twins causes them to be treated alike and to see themselves as alike. In many ways, their similarities are reinforced: Often they are given names that begin with the same letter, are given the same hair style, or are dressed identically. Thus, whatever genetic similarities exist become the cause of enforced environmental similarities that spill over onto other traits. Physical traits, such as height or cephalic index, are less susceptible to this bias. But if arm length and foot length are influenced by sports activity or by shoe style, these traits may be more environmentally correlated in monozygotic than in dizygotic twins. So-called "personality" measures will suffer greatly from this confounding. As a result, we really do not know what the heritabilities of these traits are, or even if they are heritable at all. Without random adoption studies, we cannot know how much of the observed familiality of a given trait is a consequence of common genes and how much is a consequence of common environment. In the next chapter, we take a closer look at the explanation of a trait that has received a great deal of attention from human geneticists, psychologists, and social theorists, but about which we know much less than is asserted: intelligence.

How much of the similarity of hair style, moustache, and chosen career of the identical triplets is a consequence of being treated identically from birth?

Normality and Abnormality

The difference in height between a woman who is 5 feet 10 inches and one who is 5 feet 11 inches is part of what we consider the normal variation among human beings. But is a woman who is 4 feet 6 inches (or 6 feet 6 inches) to be considered normal in a European population? Is someone with influenza normal? There is a confusion in our meanings of *normal*. Sometimes we mean *usual*, or within the range that encompasses nearly everyone. At other times, we mean *healthy*, or well adapted to the situation in which we find ourselves. Moreover, it is our usual view that the two meanings of *normal* are connected— that most people, most of the time, are in an adaptive state of health.

The concept of normality as health is not in itself independent of the concept of normality as what is usual. Human beings cannot manufacture enough of the amino acid lysine to be free of the necessity of eating it, but many bacteria can. We would clearly be better off if we could make lysine, but we do not regard it as a sign of universal disease that we are biochemically deficient in that way. Notions of functional normality are clearly derived from perceptions of commonness and rarity. But each of us has been sick at one time or another, and virtually everyone has had influenza. In fact, a rather large part of the population suffers from one chronic abnormality or another, which for them is "normal" (in the sense of being an everyday occurrence). There are whole populations in which certain functional abnormalities are very common. It is not obvious, however, whether the mild anemia of sickle-cell heterozygotes can really be thought of as an abnormality when 35% of the people in a West African population have it.

At the genetic level, everyone is heterozygous for a number of abnormal mutations—abnormal both because each of them is rare in the population and because each is deleterious physiologically and developmentally in the homozygous condition. The most recent catalogue of inherited metabolic and developmental disorders in humans lists more than a thousand that appear to be the consequence of single-gene mutations. Of these, 120 have been associated with known enzyme deficiencies. Yet all of these single-gene effects account for only a small part of the "abnormality" that is part of every human population. Most abnormalities appear to be a consequence of the genetic and environmental variation that is characteristic of quantitative traits in general.

Part of the difficulty in understanding the sources of such variation is in deciding on a model of abnormality itself. On the one hand, we might regard people whose cholesterol level is 250 milligrams per 100 milliliters of blood serum or higher as suffering from an abnormality—hypercholesterolemia—and try to analyze the genetic and environmental causes for their being in that abnormal category. On the other hand, we might simply regard such people as being near one extreme of a continuum of cholesterol levels and try to analyze the causes of variation in cholesterol levels in general. Underlying these two views are two different models of causation. The first assumes that there is one set of phenomena responsible for "normal" variation in cholesterol levels and another quite distinct set of phenomena responsible for the "abnormal" phenotype. The second assumes a single causal chain producing a distribution in which some people are, by bad luck, at one extreme.

The same issue arises when we consider mental deficiency. Is a person with an IQ of 50 to be considered within the normal range of variation in mental ability, or are we to assume the existence of a mental defect and to search for its cause? Similarly, a disease such as diabetes mellitus is not really a discrete disease with clear-cut diagnostic criteria. What is called diabetes varies from a consistent slight elevation of blood sugar to the complete inability to synthesize insulin. One consequence of this variable definition of diabetes is that almost any model for the inheritance of diabetes is as consistent with the observations as any other. Diabetes is clearly familial. If one member of a monozygotic twin pair has diabetes, the probability is 47% that the other has it, while the probability of concordance for dizygotic twins is only 10%—at least as diabetes is defined in the Danish government's registry of twins. There is no agreement about the genetic significance of the familial pattern, and recently it has been suggested that diabetes is a viral disease that is transmitted in families by infection. There

Comparison.

is a good precedent for this suggestion, one that show how dangerous it is to reason from familiality to heritability.

In certain villages in New Guinea, there is a very high incidence of a severe nervous disorder, *kuru,* sometimes called "laughing disease" because of the grimaces into which its victims' faces are distorted. The disease is progressive and invariably fatal. It is also highly familial and, for a long time, it was thought to be caused by a gene, although why such a deleterious gene should be so common in certain tribes was a mystery. It was then discovered by Carlton Gajdusek that *kuru* is, in fact, viral in origin and that it runs in families because of ritual cannibalism: When a person dies, members of his family eat bits of his brain as a way of partaking of his spirit and acquiring his virtues. But they acquire his *kuru* virus as well.

Even when there is general agreement that a distinct clinical syndrome exists, it does not follow that a special genetic or environmental cause exists that differentiates the sick from the well. There may be some underlying variable—say, an enzyme activity—that is continuously varying but that causes a distinct physiological state to appear only if it surpasses a certain threshhold. Such a model is certainly descriptive of various normal functions: Birth, menstruation, and sleep—each is a clear-cut change in physiological state that occurs when some underlying continuous variable reaches a critical level. Using data from the Danish twin registry, L. L. Cavalli-Sforza and W. F. Bodmer have estimated a kind of heritability for a variety of diseases using such a threshhold model. Their results, shown in the table below, are based on the *concordance* between the two members of a twin pair. Twins are concordant when both have or both lack a given trait.

Twin concordance and an estimate of "genetic determination," a kind of heritability, for several diseases

Disease	Concordance (%)		Estimate of Proportion of Genetic Determinism
	Dizygotic	Monozygotic	
Cancer at any site	6.8	2.6	0.23–0.33
Cancer at same site	15.9	12.9	0.10–0.15
Arterial hypertension	25.0	6.6	0.53–0.62
Mental deficiency	67.0	0.0	1.00
Manic-depressive psychosis	67.0	5.0	1.04–1.05
Death from acute infection	7.9	8.8	−0.06
Tuberculosis	37.2	15.3	0.53–0.65
Rheumatic fever	20.2	6.1	0.47–0.55
Rheumatoid arthritis	34.0	7.1	0.63–0.74
Bronchial asthma	47.0	24.0	0.58–0.71

Source: L. L. Cavalli-Sforza and W. Bodmer, *The Genetics of Human Populations* (W. H. Freeman and Company, 1971).

There is no reason, of course, why part of the variarion in "abnormal" physiology might not be the consequence of continuous "normal" variation in the population while part is the result of a traumatic cause. Elevated cholesterol levels are an example. There is a form of high cholesterol disease, familial hypercholesterolemia, that is the consequence of a single partly recessive gene mutation leading to a known enzymatic defect. Normal homozygotes have about 175–200 milligrams of cholesterol per 100 milliliters of blood serum, heterozygotes have about 300 milligrams, and homozygous abnormals have about 700 milligrams. Very few homozygous abnormals live longer than 30 years. Among "normal" homozygotes, however, there is a great deal of variation in cholesterol levels, and many people from families without the abnormal gene have cholesterol levels that are in the "abnormal" range.

Finally, we must remind ourselves that familiality of a disease is no key to its cause. When the Pellagra Commission investigated that vitamin-deficiency disease in the southern United States in 1900, it came to the conclusion that it was genetic because it ran in families. Apparently, no one on the Commission realized that poverty and malnutrition run in families too.

Variation as a Social Product

Much of our consciousness of differences between geographical and ethnic groups comes from the diversity of artifacts produced by different societies. Nothing is a more striking manifestation of that diversity than the personal adornment of different groups, as shown in these pictures. Dress, and especially ceremonial and "fancy" dress, is filled with significance both for its wearers and for its observers. Wealth, status, personal vanity, regional identity, occupation—all are signaled consciously or near-consciously by our clothes, hair styles, and jewelry. In each case, whether it is eagle feathers, rare cowrie shells, gold jewelry, or great quantities of gauzy white drapery, what is rare or magical or simply demanding of great labor in a particular society becomes the symbol to be displayed. Stripped of their finery, there is remarkably little difference between the Indian maharanee and the Polynesian princesses, between the Indian chief and the Greek countryman.

Mental Traits

7

One of the most obvious facts of our social existence is the immense variation in status, wealth, and power that exists among individuals and groups. Some people have a lot of money, while others have little; some have power over the condition of their own lives and over the lives of others, while most are relatively powerless. In all advanced countries of the capitalist world, the poorest 20% of families have about 5% of the total income, while the richest 5% have 25% of the income. Nor have the proportions changed significantly in the past 50 years. If we consider wealth rather than income, the distribution is much more asymmetrical. About 2% of the population of the United States own 25% of the wealth. But, if we exclude from consideration such commonly held property as cars and houses, that 2% own a much greater fraction (75% of the corporate stock, for example). In addition to individual variation in wealth and power, there is marked differentiation by race. The median family income of blacks in the United States is only 60% that of whites, but their infant mortality rate is 1.8 times as high, and their life expectancy is 10% shorter, as was true 50 years ago.

An outstanding feature of status, wealth, and power is that they run in families. The children of oil magnates tend to own banks, whereas the children of oil workers tend to be in debt to those banks. It was exceedingly unlikely that Nelson Rockefeller would have spent his life pumping gas in a Standard Oil station. Certainly, there is social mobility in our society, but rather less than is celebrated in song and story. The best-known study of social mobility in the United States showed that 71% of the sons of white-collar workers were themselves white-collar, whereas 62% of the sons of blue-collar workers remained in that category. These figures vastly overestimate the amount of mobility in status, wealth, and power, however, because most of the movement between white- and blue-collar jobs is horizontal with respect to income, status, control of working conditions, and security. Clerks are no less workers because they sit at desks rather than stand at benches, and salesclerks, who constitute one of the largest "white-collar" occupational groups, are among the lowest paid and least secure of all workers. Of American business leaders in 1952, 83% had fathers who were either businessmen or professionals, 10% *more* than in the first quarter of the century when farm families contributed significant numbers of children to upward social mobility.

The fact that there is familial variation in status, wealth, and power in our society is deeply troubling to many—perhaps most—Americans. We are the beneficiaries of social revolutions, extending over the seventeenth and eighteenth centuries, that were supposed to abolish inherited wealth and power. The founding fathers called for "liberty, equality, and fraternity" and assured us that "all men are created equal." Of course, they meant literally *men*—women's

Some people have a lot, some only a little.

suffrage did not come until the twentieth century—but they did not mean literally *all* men: Slavery persisted in the United States (and in British and French dominions) until the middle of the nineteenth century. One cannot make a revolution, however, with the slogan "liberty and equality for some"; so the notion that we really are all born free and equal is the cornerstone of our national ideology.

How are we to reconcile the manifest contradiction between the ideology of equality and the fact of inequality? On the one hand, one might claim that the inequalities that have characterized our society since the eighteenth century constitute a structural property of social relations themselves, that we do not really live in a community designed to give equal psychic and material benefits to all its members but that, on the contrary, our social system is built on inequality. That is, we might claim that the Declaration of the Rights of Man and the Declaration of Independence were propaganda designed to legitimize the ascent to power of a new aristocracy, the aristocracy of money. Needless to say, that point of view has not been very popular.

An alternative explanation—the one that has been dominant for two centuries—is that our society is as equal as it can be, given the natural inequalities among people. According to this view, the political and social revolutions of the eighteenth and nineteenth centuries destroyed artificial hierarchies and allowed the natural differences in ability to manifest themselves: Ours is an equal-opportunity society in which everyone starts the race of life together and with

Oliver Twist and the Artful Dodger.

the same social opportunities, but some are simply faster runners than others. It is not sufficient, however, to assert that there are intrinsic differences in ability, for that alone would not account for the passage of social power from one generation to the next. It must also be claimed that the differences are biologically inherited—that, for example, the Rockefellers of the present generation are rich not because they inherited money and power, like aristocrats of the age of Louis XVI, but because they inherited the ability to acquire money and power.

The idea that variations in intelligence, morals, gentility, and acumen are biologically inherited was a prominent theme of nineteenth-century literature. Dickens and Eliot were its greatest exponents in English. Oliver Twist, it will be remembered, was raised *from birth* in a parish work house, the most vicious and degraded social institution of the nineteenth century, where, together with "twenty or thirty other juvenile offenders against the poor-laws, [he] rolled about the floor all day, without much inconvenience of too much food or too much clothing." Yet, from earliest youth, he is the epitome of gentleness, honesty, and morality, not to mention perfect English grammar and pronunciation. In all this he contrasts sharply with young Jack Dawkins, the Artful Dodger, a person of similar upbringing who is as low and cunning a specimen of lower-class English ragamuffin as can be imagined. The reason for their difference, which is the central mystery of the novel, is that Oliver is of upper middle-class parentage: His life story is the perfect adoption study showing that blood will out.

A more extreme example is George Eliot's hero Daniel Deronda who, as the adopted son of an English baronet, spends his time in gaming and the other idle but genteel pursuits that were characteristic of young men of his class. But, mysteriously, at about the age of 21, he develops an interest in Hebrew philosophy and falls in love with a Jewish girl. The reader is not too surprised to learn, at the end of the book, that Daniel is really the son of a famous Jewish actress. Nor was this only an English preoccupation: The most widely read French authors of the late-nineteenth century, Eugene Sue and Emile Zola, used the same themes. Zola's entire cycle of Rougon-Macquart novels was explicitly designed to show the determining power of heredity over social differences.

In the twentieth century, the claims for the dominance of heredity in human affairs have been expressed less in literature and more in science. In 1905, in a scientific paper on twins, E. L. Thorndike, who was unquestionably the leading American psychologist of the day, declared that, "in the actual race of life, which is not to get ahead, but to get ahead of somebody, the chief determining factor is heredity." The firm scientific basis for this dictum may be judged from the fact that it was written a mere 5 years after the rediscovery of Mendel's paper, but 5 years before the chromosome theory of inheritance, 10 years before the development of the statistical theory of correlation coefficients, and 13 years before the foundation of the theory of inheritance of quantitative traits. In the three-quarters of a century that have since passed, the central effort of human behavioral and psychological genetics has been to put a firm foundation under Thorndike's claim.

The supposed history of the Kallikak family is often cited in psychology textbooks as an example of the critical role of heredity in determining mental and moral traits.

Martin Kallikak

He dallied with a feeble-minded tavern girl.

He married a worthy Quakeress.

She bore a son known as "Old Horror," who had ten children.

She bore seven upright, worthy children.

From Old Horror's ten children came hundreds of the lowest types of human beings.

From these seven worthy children came hundreds of the highest types of human beings.

Intelligence Testing

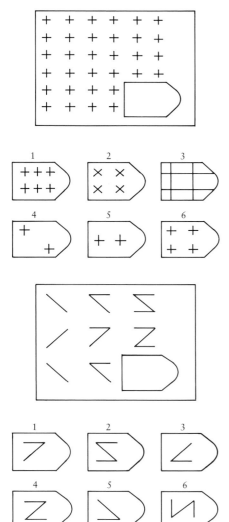

Successively more difficult progressive matrixes—a common form of nonverbal IQ tests. The object is to supply the missing form according to the logic of the forms provided.

Although there have been many studies of temperament, alcoholism, mental disease, spatial perception, and other mental characteristics, the core of human psychological genetics has been the problem of "mental ability." This concentration arises from the widespread conviction that social success in modern industrial society depends increasingly on the power of abstract reasoning. In this view, cognitive ability determines the order at the finish line in Thorndike's "race of life," an ability that must be inherited. Concentration on intelligence has been made possible technically by the creation of instruments that are said to measure differences in cognitive ability: so-called intelligence, or IQ, tests.

The IQ test was first introduced in France in 1903 by Alfred Binet in an attempt to identify those children who were not doing well in school but who would benefit from extra remedial work. The test emphasized memory, vocabulary, and the ability to discriminate among related items. This test was later modified in the United States by L. N. Terman to produce the Stanford-Binet IQ test, which became (and still is) the standard against which subsequent tests have been validated. In adapting Binet's test, Terman (and the proponents of the mental-testing movement in general) wrought a subtle but fundamental change in purpose. From being a test for singling out children who could profit from remedial work in school, the IQ test became a method for arraying all children on a scale of *intrinsic* mental ability, which was presumed to be independent of schooling and experience. The belief that IQ tests measure something that is intrinsic to the individual and beyond the influence of the environment is not an incidental feature of IQ testing; rather, it is basic to it. The very name IQ (intelligence quotient) is derived from the operation of dividing the actual scores on the test by a correction factor for age, thus restandardizing the test for each age group and cancelling out the major developmental change that occurs in mental functioning. Those who have worked on developing the tests have also cancelled out the effects of sex by weeding out those questions on which boys and girls, on the average, score differently. The tests are claimed to be without cultural bias and, in some earlier tests, to be without any linguistic bias either. This is patently untrue for verbal tests, but a considerable effort has gone into constructing nonverbal mental tests to cope with the problem of culture. The tests have not, however, been restandardized to cancel out class or race differences, inasmuch as these are the very differences the tests are meant to reveal. That is, if differential intelligence is the cause of differential social success, then a test that claims to measure intelligence had better discriminate between individuals with a larger probability of social success and those whose chance of making it is small.

A great deal of attention has been paid to the supposed fixity of IQ as opposed to the development of abilities during a person's life history. For the same person, the scores on an IQ test that is repeated within a short time are highly correlated ($r = +.95$); so the test is said to be reliable. Tests given increasing numbers of years apart become more and more poorly correlated, especially if the second test is given well into adulthood, but the correlation of tests 10 years apart is reasonably good ($r = +.80$). The different components of an IQ test, such as vocabulary, analogies, pattern recognition, and so on, are also reasonably well correlated with each other, as are different tests, including both verbal and nonverbal ones. This correlation between tests and parts of tests is regarded as a demonstration that they all measure some underlying general intelligence, the so-called "g factor," that is reflected in various ways but is itself a fixed feature of the organism that neither develops with age nor is susceptible to environmental influence. Thus, the theoretical superstructure of mental testing has a strong commitment to a biological explanation of variation in performance. The stage is set for the demonstration of the heritability of intelligence.

What IQ Tests Measure

How do we know that a test called an intelligence test does, in fact, measure intelligence? When the first IQ test was created, it was designed so that children judged on some other ground to be intelligent would do well on it. If the test had given high marks to those children everyone "knew" to be stupid, it would have been rejected. The original IQ tests were culled and adjusted so that the scores corresponded to teachers' and psychologists' *a priori* judgments about who was intelligent and who was not. The tests were tinkered with to make them the best possible predictors of school performance.

IQ tests vary immensely in form and apparent content, but many of them contain a good deal of material that obviously depends upon social class, home environment, and quality of schooling. Children are asked to identify characters and authors from English literature ("Who was Wilkins Micawber?"); they are asked to make judgments about socially acceptable behavior ("What should you do if a child younger than you hits you?"); they are asked to conform to social stereotypes ("Which is prettier?" when given a picture of a child with Negroid features and one with doll-like European features). The "right" answers to the questions do, in fact, correlate highly with scholastic performance. On the other hand, nonverbal geometric tests correlate less well with school performance. Not surprisingly, the ability to sit for a long time concentrating on a series of apparently meaningless questions is itself a reasonable predictor of school performance.

IQ tests were not developed from some general theory of intelligence and then subsequently shown, quite independently, to predict scholastic or social success. On the contrary, they were carefully designed to be predictors of scholastic performance, and the notion that they measure some intrinsic human characteristic, intelligence, has been added on with no clear justification. Indeed, there is no general agreement even about what intelligence is, and at least one educational psychologist has defined intelligence as the quality that IQ tests measure. We do not, in fact, know whether there is normal variation in intrinsic "intelligence" because we do not know how that mysterious property is to be defined. What is clear, however, is that there is considerable variation in actual school performance and that there are short-cut tests whose scores are highly correlated with that performance. That these tests are called "intelligence" tests instead of "school-performance predictors" should not mislead anyone into accepting their implicit claims.

IQ and Success

The important social claim of mental testing is not simply that IQ tests measure intelligence but that they explain the variation in social success. The equation is simple: Variations in status, wealth, and power are the result of variation in intelligence; IQ tests measure intelligence; therefore, IQ tests predict the distribution of status, wealth, and power. But do they?

The standard measure of social success used by American sociologists is not social class, a European concept whose current validity is denied by most English-speaking sociologists, but socioeconomic status (SES, for short). This is a numerical score compounded of the income, occupation, and years of schooling of the male head of household. The observed correlation between a man's childhood IQ score and his adult success, measured either as SES or as income alone, is reasonably high in most studies, about .85. Thus, IQ seems to be a good predictor of social success. There is a problem, however: Economic and social success may have many causes, including intelligence, and these causes are themselves causally linked to each other. The apparent correlation between income and IQ might be purely an indirect effect of the other causes. For example, suppose that family background were both the direct cause of good performance on an IQ test and the direct cause of success in later life. Then IQ would have a strong correlation with later success, not because IQ caused later success but because both IQ score and later success were effects of the same underlying cause. family background.

The case of IQ and success is an example of a general problem in the analysis of causes of variation. Whenever there are multiple and complex paths of causation, the simple observation that two variables are correlated does not identify the paths of causation. The figure on the facing page is a very much simplified set of possible paths of causation relating socioeconomic background, schooling, IQ, genes, and adult socioeconomic status. An observed correlation between any two variables in the figure is evidence only that there are one or more

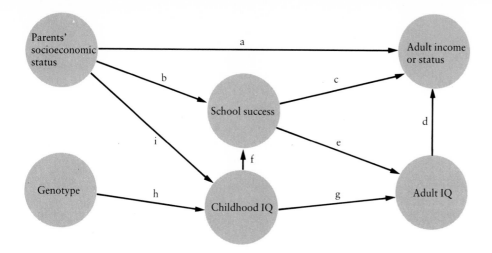

A simple scheme of possible causal paths connecting family background, genotype, IQ, and social success.

paths connecting the variables. It is not evidence of the direction of the causation or of how many steps the path contains. Thus, childhood IQ might be correlated with adult income (1) because IQ is a cause of school success, which, in turn, is a cause of income (path *f-c*); or (2) because it is a predictor of adult IQ, which is a cause of income (path *g-d*); or (3) because it is a cause of school success, which is a predictor of adult IQ, which is a cause of income (path *f-e-d*). All of these paths make IQ itself a cause of adult success. Suppose, however, that all the paths were abolished except *a* and *i*. Childhood IQ would still be correlated with adult success, but without being a cause of that success. On the contrary, it would be an effect of socioeconomic status rather than its cause.

Clearly, if we wish to understand the causes of social power, simple correlations are not sufficient. We need to look at the individual links in the scheme of causation. This can be done by examining each variable alone while holding the others constant. Thus, we can ask, "How much of the variation in adult income is predicted by variations in childhood IQ if we consider only people with the same schooling and the same socioeconomic background?" Conversely, we could hold IQ constant and see how much variation in success is explained by variation in parental success. When this was done by economists Sam Bowles and Valerie Nelson, the results were quite dramatic. The first graph on the next page shows the probability of a man being in the top one-fifth of the population in income given various amounts of schooling. Schooling here is measured not in absolute years but in what proportion of the population had as many or fewer years in school. For example, a man who fell in the lowest 10% of the population in number of years of school had only a 3.5% chance of being in the top income group, whereas a man who fell in the highest 10% of schooling had a 45.9% chance of being at the top of the income distribution. Is this because high IQ causes both school success and economic success? No. The solid bars in the graph show the probabilities when we consider only people with IQ scores near 100, which is the average for the population as a whole. There is virtually no difference. Even a man with an IQ of 100 was ten times more likely to

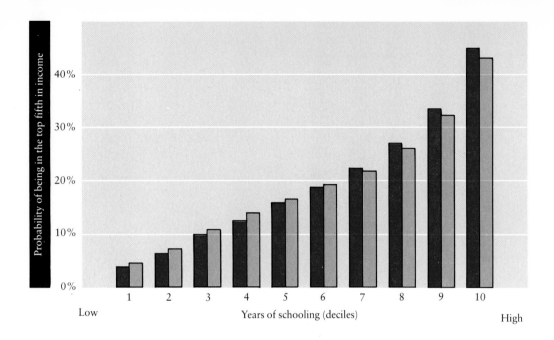

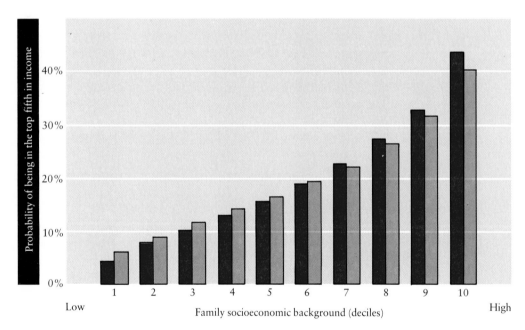

Relation between the probability of high income for the population as a whole (light-color bars) and for only persons having average IQs (dark-color bars) and years of schooling (upper graph) or social background (lower graph).

receive a high income if he was in the top 10% of schooling than if he were in the bottom 10%. Holding IQ constant does very little to the relation between years of schooling and eventual success. The second graph shows a similar comparison when family socioeconomic score rather than schooling is considered. Men whose fathers were in the top 10% of the social hierarchy were ten times more likely to receive high incomes than those who came from the poorest social stratum (43.9% compared with 4.2%). This changes only a little if we consider only men with average IQs. As the solid bars show, a man of average IQ from an upper-class family had an advantage of seven and one-half times over a man of the same IQ from the poorest family. If there is some intrinsic quality that differentiates the successful from the unsuccessful, IQ tests have failed to capture it. If such tests really do measure intrinsic intelligence, as they are claimed to do, then one can only conclude that it is better to be born rich than smart.

IQ and Genes

To postulate an inborn and unchanging basic intelligence is not the same as postulating genes for intelligence. The relation between the properties of being unborn, unchanging, and genetic are more complex than they appear. First, *inborn* does not mean *genetic*. Of the physical and physiological differences among individual people that are present from birth, many are caused not by genetic differences but by developmental noise. Small accidental alterations in the growth patterns of nerve connections in the fetal brain may produce considerable differences in mental functioning. Second, *genetic* does not mean *unchanging*, as this book has repeatedly stressed. Gene action is directly responsive to environmental signals, and the complex development and metabolism of the whole organism put it into constant interaction with the external world. Third, *inborn* does not mean *unchanging*. "Blue babies" born with an anatomical defect of the circulation can be made quite normal by a straightforward operative procedure to close off the connection in the blood supply that should have been closed off naturally during fetal development. Finally, differences between people may be unchanging without being either inborn or genetic, as those who have lost limbs, sight, or hearing in accidents can testify.

These facts have not been understood by psychologists, who have usually assumed that, if intelligence was really intrinsic, it must be genetic and that, if it was genetic, it must be unchanging. An example of this misunderstanding is a famous article by the educational psychologist Arthur Jensen, which posed in its title the question "How much can we boost IQ and scholastic achievement?" and concluded "not much" on the grounds that IQ is largely hereditary. To complete the circle of confusion, part of the evidence offered by Jensen in support of the notion that IQ is hereditary was its supposed constancy over a person's lifetime.

Estimating the Heritability of IQ

IQ scores are distributed as shown in the graph on the next page in each population for which the test was designed. The mean score is 100, and the standard deviation is 15 points. The distribution is symmetrical around the mean and has a bell shape called the *normal distribution*. There is nothing particularly revealing in any of these characteristics of the distribution of IQ scores, because the

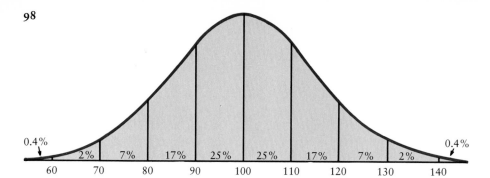

0.4% 2% 7% 17% 25% 25% 17% 7% 2% 0.4%

60   70   80   90   100   110   120   130   140

Distribution of IQ scores in a population for which the test has been standardized. The percentage of the population falling within each range of scores is given above the horizontal axis; so 17% of the people have an IQ between 80 and 90.

tests were designed and the scoring system was adjusted in order to produce a normal distribution of scores with a mean of 100 and a standard deviation of 15. When the Japanese version of the Wechsler Intelligence Scale for Children was developed, for example, it was carefully tailored to produce that distribution among Japanese school children. In the belief that important issues of social practice would be determined by knowing the heritability of IQ scores, psychological geneticists have made a considerable effort to partition the variance of the IQ distribution into genetic and environmental fractions and to establish a heritability ratio. Quite aside from the question of what use such information really is to social policy, the problems of the estimation itself are enormous.

The question, as always, is to separate genetic from environmental similarity in families. The figure on the facing page gives observed correlations in IQ scores between various kinds of related and unrelated people. Both medians and ranges over various studies are given, together with the correlation expected if the heritability were 100% and the effects of genes on IQ were additive. The more closely related the people, the higher the correlation in their IQ scores. The median correlations between parent and child, between siblings, and between dizygotic twins are all close to the simple genetic expectation of .50. Unrelated people have a much lower correlation, and identical twins have a much higher one. An increasing correlation with increasing family relationship is predicted by any theory of the causation of IQ, however; so the gross pattern is not very informative.

The very large ranges for each class of relationship are disturbing. It is difficult to have too much confidence in the studies of parent-child correlation when the results are evenly spread over a range of $r = .20$ to $r = .80$, or of the siblings who are, again, remarkably evenly spread between $r = .30$ and $r = .80$. Under the circumstances, the fact that their median values fall close to .50 seems more like a numerical artifact than the revelation of any biological reality. These ranges correspond to a range of heritability between $H^2 = 40\%$ and $H^2 = 160\%$—not a very reassuring result.

Comparisons between people with the same genetic relationships but different environmental circumstances and those between people with different genetic relationships but the same environmental circumstances have been most

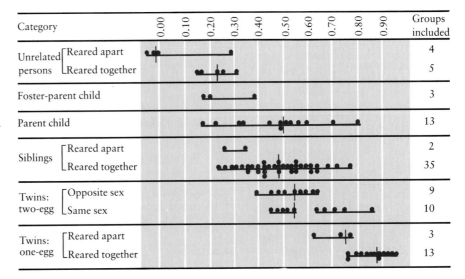

Category	0.00	0.10	0.20	0.30	0.40	0.50	0.60	0.70	0.80	0.90	Groups included
Unrelated persons — Reared apart											4
Unrelated persons — Reared together											5
Foster-parent child											3
Parent child											13
Siblings — Reared apart											2
Siblings — Reared together											35
Twins: two-egg — Opposite sex											9
Twins: two-egg — Same sex											10
Twins: one-egg — Reared apart											3
Twins: one-egg — Reared together											13

Observed correlations in IQ between people of different degrees of relationship. Each point is a separate study and each vertical line shows the median value for each degree of relationship.

often used to estimate the heritability of IQ. The IQ scores of unrelated people are uncorrelated when they are reared apart, as indeed they must be under any theory of causation, but their scores have a median correlation of about .25 when they are brought up together. One study produced a correlation of .30 for the IQ scores of unrelated persons raised apart, which should caution us against accepting scientific results too readily. There is no way that a truly random sample of unrelated persons can be correlated in their IQ scores. Either the study was badly designed or the result was a statistical fluke. In either case, it gives one pause. Comparing the correlation between the IQs of foster parents and those of their foster children with the correlation between the IQs of biological parents and those of their children again show the effect of a common environment. At the other end of the scale, the IQ scores of identical twins reared together have a median correlation of .88, whereas those of identical twins reared apart have a correlation of .75. Again, the effect of a shared family environment is detected.

If we take the results in the figure above at face value for the moment, we can see that there are a variety of ways in which to estimate the heritability of IQ, and they all agree in giving a fairly high value. The .25 correlation between unrelated persons reared together might be regarded as a direct estimate of the effect of a common environment. Monozygotic twins raised together have both genes and environment in common, and their median correlation is .87. So, correcting for the effects of a common environment, we can get an estimate of heritability of .87 − .25 = .62. Alternatively, we can use the standard comparison between dizygotic and monozygotic twins. The median correlation for monozygotic twins is .87 and that for dizygotic twins is .53. Therefore,

$$H^2 = \frac{r_M - r_D}{1 - r_D} = \frac{.87 - .53}{1 - .53} = .72.$$

However, the trouble is that we cannot take any of these studies or theories at face value. We have already discussed (p. 79) the dubiousness of the assumption that identical and fraternal twins are really treated in the same way; so the estimate of .72 is too large by some undetermined amount. This problem also contaminates the first estimate. Can we really assume that the correlation of .25 between the IQ scores of unrelated persons reared together is also an adequate estimate of the strength of the effect of a common environment on identical twins? If not, then we have undercorrected for common environment in the first estimate as well. Indeed, we ought to be suspicious of hanging too much generality on any study of what is an extraordinary human relationship.

Twins Raised Apart

Twins separated from birth have a fatal fascination for the human geneticist, just as they do for the romantic novelist. Alexandre Dumas's *Corsican Brothers*, separated by a knife at birth but feeling simultaneous pain and pleasure over a distance of many miles, are but the literary counterparts of Sir Cyril Burt's no less fictitious identical twins who have identical IQs although they have never seen each other. The scenario seems perfect for an examination of the genetic effects on human variation: Identical twins have identical genes; if, raised in unrelated environments, they nevertheless show similarities in their tests, these similarities must be genetic. In fact, the correlation between the IQ scores of identical twins raised apart is a direct estimate of the heritability of IQ. As the figure on page 99 shows, the estimate is .75.

A little rumination on the subject of identical twins raised apart begins to raise questions. Nineteenth-century novels aside, what do we imagine the circumstances to be that will separate identical twins in earliest infancy? Are the twins really raised in utterly different environments? There are only four studies of IQ in separated twins reported in the literature on this subject, and these have been the subject of a detailed analysis and review, by the psychologist Leo Kamin, that produced disquieting results. As might be expected, none of the studies contains many twin pairs: Juel-Nielsen, 12 pairs; Newman et al., 19 pairs; Shields, 44 pairs; and Burt, 53 pairs. Newman and Shields got their pairs by advertising in newspapers or on television and then culling the respondents' mailed replies. Thus, it seems there was self-selection on the part of the respondents, each of whom was sufficiently similar to the other twin to regard the pair as identical and each of whom was in contact with his or her "separated" twin.

Only Shields provided detailed life histories for his twins, and it turns out that they were not really separated at all. In real life, twins are separated at birth because the mother has died, or because the parents cannot afford to keep them both, or because they are too sick to do so. The children are typically given to aunts, sisters, or best friends, and are brought up in neighboring houses in the same towns. In Shields's study, every twin pair but 4 was raised by close relatives, close friends, or neighbors. Twin pairs separated at birth and raised in unrelated environments belong more to the realm of romance than to reality.

The largest and most widely used study of separated twins is actually a series of studies presumably carried out by Sir Cyril Burt and his collaborators over a period of 20 years. The researchers maintained that there was no significant correlation in economic status between the families raising the separated twins. No details were given. When Kamin examined the studies carefully, curious anomalies appeared. Sample sizes were reported differently, or sometimes not at all, in different reports. No details were given about the IQ tests. Test scores were adjusted to account for the interviewers' subjective perception of whether they adequately reflected the similarities of the twins. Correlation coefficients computed on different sets of twins nevertheless agreed over and over to the third decimal place. The original data, moreover, were supposed to have been lost in a laboratory fire.

Several years of investigation by Kamin and, later, by the medical writer Oliver Gillies finally revealed that Burt's twin studies were a complete fabrication. Burt's named collaborators did not exist, the test scores did not exist, and, it appears, the twins did not exist either. For this reason, all studies by Burt and his "collaborators" were left out of the figure on page 99. Burt apparently had a long history of fabrication, including laudatory reviews of his own work, published under fictitious names in a journal of which he was the editor.

The reactions of psychologists and human geneticists to the succession of revelations about Burt's work are in themselves revealing. Some said that Burt had merely "carelessly reported" his work. One can be careless, it seems, to the third decimal place. Others said it was a phenomenon of Burt's old age, a result of senility, but it became clear that the frauds went back to his early work as well. Most disturbing of all, some of his senior colleagues said that they had always doubted Burt's reports but that they had never challenged him because he "said it with such style."

Burt's fraudulent reports and the reactions of his colleagues to them are only the extreme of a general phenomenon in the study of human mental and temperamental variation. Most studies of the heritability of mental traits are marked by one or more serious methodological defects, including (1) very small sample size; (2) confusion of observed correlation between relatives with genetic correlation; (3) selective adoptions in fostering studies; (4) subjective ratings of similarities; and (5) after-the-fact statistical adjustments that bring the data more in line with genetic expectations. Any or all of these defects would automatically disqualify a research report for publication in a scientific journal if the subject were milk yield in cattle. Journals of psychology and behavioral genetics regularly publish them, and no progress in the rigor of such work is apparent. As late as 1979, the major scientific journal of behavioral genetics published an estimate of the heritability of human IQ based entirely on the observed correlation between parents and offspring in normally structured families, even though the editor knew, and has elsewhere stated, that there is no way of knowing in such cases how much of the correlation is a consequence of shared family environment. It is impossible to avoid the conclusion that there is a deeply established prejudice in favor of a genetical explanation of human behavioral variation.

Adoption Studies

In principle, it should be possible to assess genetic influence on IQ variation from adoption studies. The figure on page 99 includes data from studies comparing either the IQs of parents with those of their foster children or the IQs of siblings raised apart. On the face of it, the two sets of studies give very similar results, although the expected genetic correlation between the IQs of siblings raised apart is .5 and that between the IQs of foster parents and those of their foster children is zero. Thus, there is not much evidence here for heritability.

The ideal adoption study would compare the IQ of an adopted child with the IQs of its adopted parents and those of its biological parents. This ideal is difficult to realize, however, because it is usually impossible to obtain the data about the biological parents. A substitute is to use a different group of children and their biological parents, but to try to match the characteristics of the biological and adoptive families as closely as possible. In either case, it is vital that the adoptions have been made at random. That is, it is essential that children of parents whose IQs are high are not fostered into families with higher than average IQs. Otherwise, an IQ correlation may appear between foster children and their biological parents that may be a consequence not of the biological relation but of the environment in which they were raised.

There have been three large adoption studies that have used this comparative design. Two, the studies by Burks and Leahy, compared foster families with a different, but supposedly matched, set of biological families. The third, the study by Skodak and Skeels, used educational levels of the foster mothers, IQs of the children, and IQs of the biological mothers of the foster children. Unfortunately, no IQ tests were given to the adoptive mothers. The results shown in the table on the facing page seem very strong evidence for genetic effects, inasmuch as the correlation of children with their foster parents in each case is so

Observed correlations in IQ or educational attainment between parents and children in three studies of adoption	Correlations		
Study	Foster Child with Foster Mother	Biological Child with Biological Mother	Foster Child with Biological Mother
Burks	.19	.46	—
Leahy	.20	.51	—
Skodak and Skeels	.02	—	.32

much lower than the correlation of the children with their biological parents. A careful examination of these studies by Kamin, however, raises some serious questions about their design. The study by Burks included many severely retarded children, which, of course, greatly reduced the IQ correlation with the adoptive parents, who in general come from higher socioeconomic categories and have higher IQ scores than the average for the population as a whole. In both the Burks study and the Leahy study, the matching of biological and foster families was poor. Adoptive parents were older, had incomes that were higher by 50%, and had fewer children, as might be expected. They were, in general, much less variable than the biological families in nearly every respect. In the Skodak and Skeels study, there were selective adoptions, with children of highly educated mothers being placed in higher-status homes. So, from such studies we really do not know what the heritability of IQ is.

The most striking and consistent feature of adoption studies is not usually much commented upon by those interested in demonstrating genetic effects: Whatever the correlations may be between the IQs of children and those of their biological parents, *the phenomenon of adoption raises children's IQ significantly.* In the study by Skodak and Skeels, the biological mothers' IQs averaged only 86, one standard deviation below the average for the population. By contrast, the mean IQ of their children who were raised by adoptive families was 117, one standard deviation above the mean for the population. In a study of orphanage children in the United Kingdom, the same phenomenon was seen. Children taken into the orphanage in early infancy had an average IQ of 105 if they remained in the orphanage until they were about 5 years old, 100 if they were returned to their biological mothers, but 115 if they were adopted. These observations are just what is to be expected from the social characteristics of adopting parents: They are generally middle-class and upper-middle-class couples, with few or no children of their own, who have the financial power, the motivation, and the class background to produce "intelligent" children. There is, of course, no contradiction between this adoption effect and the possibility that IQ scores may be highly heritable. "Genetic" does not mean "unchanging." No matter how high the heritability of IQ might prove to be, upper-middle-class families tend to produce upper-middle-class children.

Migration

The movement of genes from one population to another has been a constant process throughout history. Rarely, as is true of the migration of middle-class Germans to the United States in the 1920s, is it a consequence of more or less freely chosen emigration. More often, it is a consequence of political and economic forces that leave the people migrating little or no choice. The immigration of southern Europeans to the United States at the beginning of this century and the earlier influx of Asians were the results of the intolerable economic conditions in which those people found themselves at home and the active recruitment of immigrants by labor agents and steamship companies. Still others migrate when they find the political climate of their country to be hostile to their own religious, social, or economic ideology, as was the case for the Pilgrims of the seventeenth century and the Vietnamese "boat people" of the twentieth century.

Beyond the movement of people by their own acts of will, violent coercion has played a major role in molding human geographical variation. The legendary rape of the Sabine women by Romulus's male followers symbolizes the way in which local tribes and groups often intermix. On a vaster scale, the transport of enslaved Africans to the New World beginning in the sixteenth century, and continuing for 300 years, not only had a profound effect on the diversity of American populations, but was probably the most significant phenomenon in directing the political and social evolution of the western hemisphere.

And lo !. he darts his piercing eye profound,
And looks majestically stern around !

— The husband and wife, after being sold to different pur-
chasers, violently separated....never to see
each other more.

Diversity among Groups

8

Both the biology and the history of the human species have resulted in the appearance of lines of division that separate human beings into groups. There are men and women, blacks and whites, Germans and Spaniards, workers and bosses. The biological and cultural differentiation of these groups varies, but all these distinctions have two things in common: All of them have an historical contingency, even the most biological of them, and all have been claimed to be biological, even the most historical of them. It is trivially obvious that the primary distinguishing feature of men and women, the differential development of their internal and external sex organs, has a very ancient biological origin. In some considerable detail, that difference is common to all mammals and, in its more general features, of all vertebrate creatures. Yet the differences in dress, social status, role in the family, and the productive process that also characterize men and women are totally historically determined. They differ immensely from time to time and from culture to culture. It is a recurrent theme in discussions of sex differences in our society that these cultural differences are claimed to be based in biology. Recently, both the popular and the scientific press have been filled with speculations about a "math gene" possessed by men but not by women as a way of explaining the low proportion of women among mathematicians. At the other extreme of group differences are workers and entrepreneurs, categories that have no meaning at all among the Pygmies and could be analogized only with the greatest difficulty with the organization of European society in the thirteenth century. Still, there are repeated claims that the working class and middle class that have developed since the seventeenth century in Europe are, in fact, biologically based. In the words of the psychologist R. Herrnstein:

The privileged classes of the past were probably not much superior biologically to the down-trodden, which is why revolution had a fair chance of success. By removing artifical barriers between classes, society has encouraged the creation of biological barriers. When people take their natural level in society, the upper classes will, by definition, have greater capacity than the lower.

The biological and the historical do modulate each other in the creation of group differences. Plantation owners and their slaves did not come into existence in the Americas because there were whites and blacks, but because the plantation system was suitable for the production of tobacco, sugar cane, and cotton and because armies of cheap labor to produce these crops could be procured from slavers. For historical reasons, the slaves were black and the owners white, so that a biological difference—skin color—became a sign for a class difference. Subsequently, that skin-color difference became a handle for political and economic discrimination. As a result, skin color is, in fact, a cause of

unemployment. Thus, we might say that unemployment is in the genes, although by an indirect and entirely historical path of causation. Moreover, biology can feed back onto biology through social distinctions: For hormonal reasons, women, on the average (but only on the average), have a different proportion of muscle to fat than men, and this has the consequence that women, on the average (but only on the average), can exert somewhat less physical force on objects. The division of labor between men and women and the division of early training, activity, and attitude cause a very considerable exaggeration of this small difference, so that women become physically weaker than men during their development to an extent far in excess of what can be ascribed to hormones. Women's sports records have not equalled men's, but the differences have narrowed very considerably in the past 20 years.

Two major related problems in describing and understanding the differences between groups are assignment and stereotyping. How do we decide that person A is male and person B is female? In principle, one can always decide upon some *a priori* criterion and apply it strictly: A is male because he has a penis. There are, however, a small number of people for whom even such a clear-cut criterion will not work because their external genitalia have developed ambiguously. The number of such people is sufficiently small, however, that this does not pose a serious problem for classification. The real problem is that lines are drawn differently in life. A whole suite of characteristics is assigned to each group to create a complex stereotype, which we then use to judge group membership. We expect everyone to conform to those stereotypes, even if there is considerable variation within groups and overlap between groups when the characteristics are actually tabulated. Moreover, we make special note of any deviations from expected stereotypes: "What an aggressive woman!" "He runs like a girl." "What a placid temperament she has for a Latin." "His skin is so light you'd never know he was a black." If the dissonance is too great, we may become disturbed, disoriented, or even angry. In one of his novels, Paul Scott described an Indian who, having been brought up from the age of two in English upper-middle-class society and having gone to an English boarding school, returns to India only to confront, with his flawless upper-class English voice and attitudes and his dark brown face, British colonial society. The bafflement he creates among the lower civil-service workers, who know an upper-class Englishman when they hear one and an Indian when they see one, is more than they can bear.

The pervasiveness of stereotyping is so complete that it requires a special effort, even as I write these words, not to use phrases like "women are . . ." or "blacks differ from whites in. . . ." Such locutions say that "*all* women are . . ." or that "*all* blacks differ from *all* whites in. . . ." But there are no characteristics that distinguish all blacks from all whites, not even skin color. Once it is admitted that a characteristic is variable within a group and that there is some overlap of the distributions of the groups, the stereotyped description is false, and we need a more careful and subtle description of the differences between groups. The figure at the left shows the distribution of weights of young adult white men and the distribution of weights of young adult white women in the United States. There is a considerable overlap in the two distributions, but the mean male weight is greater than the mean female weight. Is there any more that can be said? It is not true, for example, that most men weigh more than the heaviest woman. Sometimes it is said that "most men weigh more than most women," but that turns out to be nothing more than the statement that the median male weight is greater than the median female weight. If two populations differ in their medians, even by the slightest amount, then, literally, most individuals in one population are above most people in the other. The critical point is that, if the distributions of a characteristic in two populations overlap each other, stereotypic statements are misleading. A person's weight cannot be predicted by knowing his or her sex, nor can sex be predicted from weight. Each person must be considered individually.

In actual social practice, groups are well defined when an important social function is served by defining them, although the marks of social definition are often, as isolated criteria, almost meaningless. In the United States, a person is usually considered black if she or he has any known black ancestry, irrespective

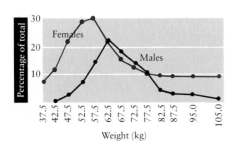

Distributions of the body weights of young adult (17-year-old) American females and males.

of actual appearance, or is usually considered a Jew if she or he has a Jewish name or Jewish parents, irrespective of religious attitudes. These definitions are social realities. The question is whether those social realities correspond to any other kind of reality, whether differences in race, class, and sex capture any significant part of human variation beyond the social critera used to define the groups in the first place.

What Is Race?

A Basque from the Pyrenees between France and Spain. These people are sometimes said to be direct descendants of early neolithic western Europeans with no admixture with other populations.

The total population of the human species is clearly differentiated along geographic lines. No one would mistake a Chinese for a West African or a Finn for an Australian aborigine. Before the development of the modern theory of evolution, when it was assumed that all species had been specially created by a supernatural power, it was also supposed that those geographic groups were the descendants of separately created "nations." Each, it was believed, represented a pure Platonic type, just as did every species, types that had remained unchanged since the moment of creation. The acceptance of the fact of evolution at the end of the nineteenth century meant the destruction of the typological view of species, inasmuch as Darwinism emphasized both the fact of individual variation within species and the constant modification that each species is undergoing. Curiously, this change in the view of biologists was not felt in anthropology until nearly a century after the publication of Darwin's *Origin of Species*. Until recently, anthropologists were fairly typological in their thinking, and textbooks of physical anthropology were much concerned with marking off the boundaries of human races and giving them names. Some of the authors ("lumpers") named only a dozen major races, while others ("splitters") distinguished scores. One widely accepted treatise listed 30 races and gave a photograph of a typical member of each.

The problem with these categories is that there are too many contradictions between the different ways of dividing human races. Are the Turks Caucasians, as their appearance suggests, or do they belong with the Mongoloid tribes of Central Asia, with whom they (along with the Hungarians and the Finns) have linguistic affinity? What is to be done with the Basques, who appear to all eyes to be Spaniards and yet whose language and culture are seemingly unrelated to any other in the world? The Hindi and Urdu speakers of India are a special problem: Historically, they are a mixture of South Asian Dravidian aborigines, central Asian Aryans (who are clearly related to Caucasians), and Persians. Should they be put together with Europeans, whose languages come from Sanskrit, to which Hindi and Urdu are very close, or do their dark skins make them South Asians?

The effort to build a more and more complex set of categories of human types that would correspond to the incredible variety of human beings finally fell

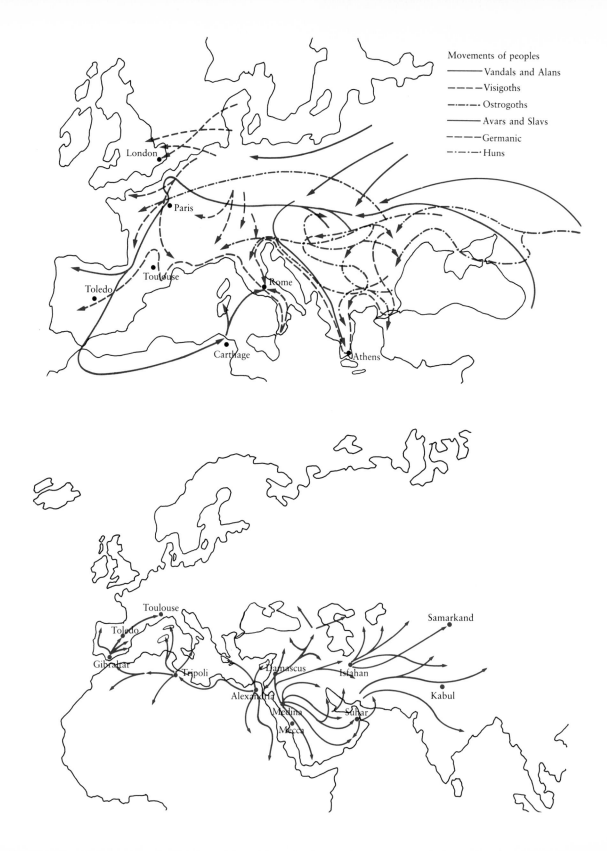

Movements of peoples
— Vandals and Alans
- - - Visigoths
-·-·- Ostrogoths
— Avars and Slavs
- - - Germanic
-·-·- Huns

London

Paris

Toledo

Toulouse

Rome

Carthage

Athens

Toulouse

Toledo

Gibraltar

Tripoli

Damascus

Alexandria

Medina

Mecca

Suhar

Isfahan

Samarkand

Kabul

The upper map reconstructs the history of European migrations and invasions of the first six centuries A.D. Avars, Huns, Alans, and Ostrogoths invaded Europe from the east, mixing with, and driving to the south and east, the previous residents of central Europe.

Beginning in 632, at the death of Mohammed, the people of Arabia and North Africa spread over the Mediterranean Basin and the Near East to India so that by 750 the Abbasid dynasty was the major civilization of the western world, as shown in the lower map.

under its own weight, just as the system of Ptolemaic epicycles did when the observations of astronomy forced it into more and more convoluted explanations. As the Copernican revolution simplified astronomical explanations, so the Darwinian revolution, when it finally reached anthropology, simplified our understanding of human geographical variation. Anthropologists no longer try to name and define races and subraces, because they recognize that there are no "pure" human groups who have existed since the Creation as separate units. The most striking feature of global human history is the incessant and widespread migration and fusion of groups from different regions. Wholesale migration is not a recent phenomenon brought about by the development of airplanes and ships; it has been an economic necessity at all times. Britons, so conscious of their race, are, in fact, an amalgam of the Beaker Folk of the Bronze Age, the Indo-European Celts of the first millenium B.C., the Angles, Saxons, Jutes, and Picts of the first millenium A.D., and, finally, the Vikings and their parvenu grandchildren, the Normans. The adjoining figures show the mixing of Central Asians, Slavs, Mediterraneans, and Northern Europeans that occurred in the short period at the end of the Roman Empire. The rise of Islam in the seventh century A.D. resulted in a widespread diffusion of Arabic culture and people from Cordoba to Kabul in less than a hundred years. The Umayyad caliphate was in Spain by 711, in Toulouse by 721, and in Corsica by the middle of the ninth century, and the last Arab state was not expelled from Europe until 1492, by their most Catholic Majesties Ferdinand and Isabella. Meanwhile, Arab traders and slavers had thoroughly penetrated central Africa. When Henry M. Stanley set out on his famous search for Dr. Livingstone, he began in the Arabized port of Zanzibar and followed a well-established Arab trade road west, through Arab trading posts, to Lake Victoria. Nor is the situation different in the Far East. The Japanese are a mixture of Korean invaders and northern islanders. Even the Australian aborigines, who were regarded as virtually another species by some anthropologists, have large infusions of Papuan and Polynesian ancestry on the eastern and northern coasts of Australia. The notion that there are stable, pure races that only now are in danger of mixing under the influence of modern industrial culture is nonsense. There may indeed be endogamous groups, largely biologically isolated by geography and culture from their neighbors, such as the Pygmies of the Ituri Forest, but these are rare and not perfectly isolated in any event. Colin Turnbull's Pygmy companion, Kenge, was reported to have had sexual encounters with women of other tribes.

An Example of Migration

The passage of genes between populations occurs under many circumstances: wholesale migration, wars, trade, slave-taking, rape, and simple propinquity. A well-documented case involving the last three is the admixture of blacks and

Estimates of proportion of European ancestry in American blacks from the frequency of the Fy^a allele

Locality	Fy^a Frequency in Blacks	Estimate of Migration
New York City	.081	.189
Detroit	.111	.260
Oakland, California	.094	.220
Charleston, South Carolina	.016	.037
Evans and Bulloch counties, Georgia	.045	.106

Source: L. L. Cavalli-Sforza and W. Bodmer, *The Genetics of Human Populations* (W. H. Freeman and Company, 1971).

whites in North America. Black slaves were introduced into continental North America from West Africa beginning in the seventeenth century. They continued to be imported until the economics of plantation life and world trade at the beginning of the nineteenth century made the cost of purchase greater than the cost of slave reproduction. From 1813 to the present, then, with the exception of some immigration from the Caribbean, America has had virtually no influx of African genes. During the entire period since the introduction of slavery, there has been a certain amount of admixture between blacks and whites, varying from time to time and from place to place. Slave women were raped or taken as mistresses by their white masters. Thomas Jefferson had several children with his slave mistress, Sally Hemings. Intermarriage between blacks and whites was prohibited in the South after the emancipation, but the admixture continued in cities, especially in industrial concentrations in the North. At present, people classified as black have variable amounts of white ancestry, and some fraction of whites have some black ancestry. Because of the social definition of race in the United States, these two categories are very asymmetrical: Almost any known black ancestry makes a person "black."

It is possible to estimate the actual proportion of white ancestry among blacks from data on genetic polymorphisms. There are a few polymorphisms in which one allele is common among Europeans but absent or virtually absent among West Africans, or vice versa. The Duffy blood group allele Fy^a has a frequency of about 40% among whites but is essentially absent among West Africans. The allele R_0 of the Rh blood-group system is rare among whites, but it has a frequency of 60% among West Africans. To the extent that black Americans have both African and European ancestry, the frequency of these alleles among them should be intermediate. If, on the average, the black population had 10% white ancestry, we would expect the frequency of an allele among them to be the result of mixing 90% of the African allele frequency with 10% of the European allele frequency. For the Fy^a allele, the frequency among American blacks would be $.10(.40) + .90(0) = 4\%$. The table at the left shows the results of several studies of allele frequencies among American blacks as compared with American whites and West Africans. The estimates run from about 4% admixture in Charleston to more than 25% in Detroit, with the rural south being intermediate. It is not clear whether these locality differences reflect actual differential rates of admixture or differential rates of migration into cities of blacks with different amounts of European ancestry. If Charleston is typical of southern cities, then, taking into account the current distribution of blacks between rural and urban North and South, about 15% of all genes in the American black population have been introduced from Europeans in the past 250 years or so. Taking 25 years as a generation, that means a rate of roughly 1.5% per genera-

Black Americans have varying amounts of African and European ancestry.

tion. However, because these calculations do not take into account the untraceable passage of black genes into the white population, they are underestimates of actual admixture rates. Once the proportion of white ancestry is high enough in a family, that line disappears from the "black" population.

Geographical Differentiation

The migrations and fusions of people have not resulted in a uniform blend. The processes of differentiation between local groups in the course of their evolution, which we will look at more closely in the last chapter, go on at the same time as the admixture, so that there is always some geographical variation. Some of this variation is a vestige of past divergence that has not yet been ironed out by migration and mating. As a result, one can still see the image of past

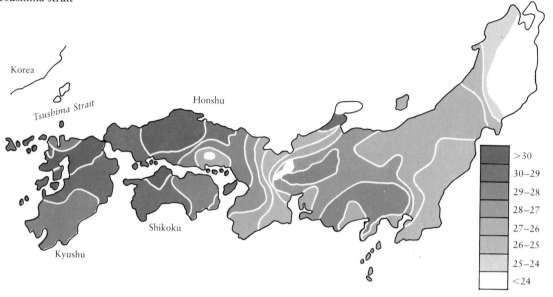

Clines in the percentage of blood group A in modern Japan, radiating out from the region just across the Tsushima strait from Korea.

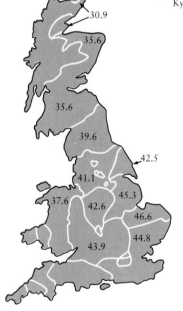

Percentage of blood group A in various regions of Britain showing a decreasing cline northward and westward from the original invasion sites of the Danes on the east coast (so-called Danelaw).

invasions in the present frequencies of genes. The figure above shows the cline (series of gradual changes) in the frequency of blood type A in Japan. Blood type A is highly concentrated in the west opposite the Korean mainland, and it decreases clinally to the northeast toward Hokkaido. Korea has been the traditional source of continental invasions of Japan, a tradition that is embodied in the story of the *kamikaze,* the Divine Wind that scattered Kublai Khan's fleet in the Tsushima Strait in 1281. Similar clines exist in Britain and Ireland. The frequency of blood type A is higher than 50% in East Anglia, but it declines toward the north and west, reaching 25% in western Ireland. Types B and AB have the opposite trend, rising from 8% in the east and south to as high as 18% in Scotland, Wales, and Ireland. These changes reflect what is known of the movements of the Celts, Picts, Danes, and Vikings through the islands.

On a broad scale, there are large and obvious differences in skin color, hair form, stature, and language from one geographical region to another. The patent differences between geographical groups seem vastly greater to us than variations from individual to individual within groups, but these differences are in traits to which we are particularly sensitive. It is those very characteristics that we use to recognize individuals and to which we are so finely attuned that appear to differentiate groups as well. The skin color of any Ghanaian is so obviously different from that of any Dane that the difference between the two groups seems vastly greater than the minor variation between the individual members of either group. Typology and stereotyping slip back into our perception of groups precisely because we are attuned to those differences between groups that turn out to be large. The question is, are these large differences typical?

To answer that question at the biological level, we need a sample of variation that can be described objectively, that is not weighted by characters that happen to be obvious to us as social observers, and that has been characterized over a sufficiently large sample of geographical groups to give us a general picture. Simple molecular polymorphisms seem to fill the bill. Variations in blood groups, immune types, and enzymes can be characterized objectively; their genetics is simple and well understood; and, in the past 40 years, anthropologists and human geneticists have tested the blood groups and enzymes of thousands of people in scores of local populations throughout the world. Comparing the gene frequencies of these various polymorphisms in different populations should give us an objective picture of human geographical diversity.

An example of such a comparison is given in the adjoining figure, which shows the frequencies of the ABO blood-group alleles in a sample of human geographical groups. Each point represents a population. The frequency of each of the three alleles $(I^A, I^B,$ and $i)$ for that population can be read from the diagram by measuring the perpendiculars to the point from each of the three sides of the triangle. Thus, the frequencies for population 11 are $I^A = .20, I^B = .28,$ and $i = .52$. A totally monomorphic population would be a point at one of the three corners. A population on one of the sides is missing one of the three alleles—for example, populations 4, 5, 14, and 18, all of which are missing allele I^B. No human population is completely monomorphic for the ABO system, although, in a sample of 194 Toba Indians from Argentina, only three people were of blood type A and everyone else was of type O. As the figure shows, all human populations are clustered in the upper right-hand corner, corresponding to a high frequency of i, an intermediate frequency of I^A, and a low to intermediate frequency of I^B. No human population is known, for example, with a frequency of the i allele that is less than 50% or with a frequency of the I^B allele that is more than 30%. The other feature of the worldwide distribution of ABO frequencies is that they do not have any strong geographical consistency. The shaded areas enclose populations with similar frequencies, yet these clusters do not correspond to geographical groups. Each cluster contains some mixture of European, African, Asian, and American Indian populations. That is not to say that there are no geographical regularities at all. The various groups of American Indians have uniformly very low frequencies of the allele I^B, and some have

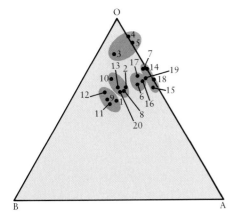

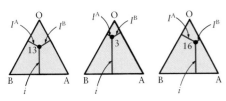

Triallelic diagram showing the ABO blood group composition of 20 different human populations from various regions of the world. Populations 1–3 are Africans, 4–7 are American Indians, 8–13 are Asians, 14–15 are Australian Aborigines, and 16–20 are Europeans.

Worldwide distribution of the blood group gene I^B. The lines connect places with equal frequencies of the gene. Note that central Asia is a region of high frequency of I^B, with the frequency decreasing in all directions.

virtually nothing but i. However, Australian aborigines and Basques also have very low frequencies of I^B, and some American Indian groups have fairly high frequencies of I^A. In fact, both the highest I^A frequency in the world (82%) and the lowest (1.5%) are found in American Indian tribes. The allele I^B is generally in its highest frequency (20%–25%) in central and southeastern Asia, and decreases in concentric zones in all directions from the central Asian plateau (see the figure above). Perhaps the I^B allele originated in central Asia and then spread with the movement of central Asian steppe people into Europe, Africa, and Siberia. The complete absence of the I^B allele in the New World and Australia suggests that its origin would have occurred after the rise in sea level that accompanied the melting of the continental glaciers 10,000 years ago.

The general impression given by the triangular diagram on page 117, that human populations are moderately differentiated from each other by gene frequencies, is borne out when a broader sample of genes is examined. The table on the facing page gives the frequencies of alleles for seven polymorphic enzyme loci chosen at random in a sample of Europeans and Africans. The similarity between the two groups is striking. Except for the phosphoglucomutase-3 gene,

Allelic frequencies at seven polymorphic
loci in Europeans and black Africans

Locus	Europeans			Africans		
	Allele 1	Allele 2	Allele 3	Allele 1	Allele 2	Allele 3
Red cell acid phosphatase	.36	.60	.04	.17	.83	—
Phosphoglucomutase-1	.77	.23	—	.79	.21	—
Phosphoglucomutase-3	.74	.26	—	.37	.63	—
Adenylate kinase	.95	.05	—	1.00	—	—
Peptidase A	.76	—	.24	.90	.10	—
Peptidase D	.99	.01	—	.95	.03	.02
Adenosine deaminase	.94	.06	—	.97	.03	—
Average heterozygosity per individual	.068 ± .028			.052 ± .023		

Source: R. C. Lewontin, *The Genetic Basis of Evolutionary Change* (Columbia University Press, 1974).

the most common allele of each gene is the same in Europeans and in Africans. There are low-frequency alleles that appear in one sample but not in the other—for example, allele 2 of the adenylate kinase gene, which is present at a low level among the Europeans but is absent from the African sample—but these might appear in larger samples. Compared with the enzyme diversity within each population, the differences between the two groups are not great. A similar comparison with very similar results is shown in the table on page 37, which lists the frequencies of the much more polymorphic HLA types.

To give some feeling for how different and how similar populations may be genetically, the table on the next page gives the three polymorphic genes for which the major human geographical groups are most different and the three for which they are most similar. The most different are very different indeed: Blacks and Asians each have a different allele in very high frequency for the Duffy blood group, and the possession of the allele Fy^a is a fairly certain indication of some European ancestry. The same asymmetry between blacks and Asians appears for the *MNS* and *P* genes as well. It is interesting that the Europeans are rather more genetically variable for all three genes than either of the other "races." Perhaps Europeans are more mixed in their ancestry than the "purer" Africans and Asians. Whatever the genetic differences among the three major groups, however, there is no case in which one group is completely homozygous for one allele while a different group is totally homozygous for a different allele. All groups share some alleles with other groups. There are no pure racial genes, although there are alleles that make their appearance in only one geographical group, like the Fy^a allele among Caucasians or the Diego blood-group factor, which is found only among Asians and American Indians, but never in very high frequency even there.

Examples of extreme differentiation and close similarity in blood group allele frequencies in three racial groups

Gene	Alleles	Caucasoid	Negroid	Mongoloid
Duffy	Fy	.0300	.9393	.0985
	Fy^a	.4208	.0607	.9015
	Fy^b	.5492	—	—
Rhesus	R_0	.0186	.7395	.0409
	R_1	.4036	.0256	.7591
	R_2	.1670	.0427	.1951
	r	.3820	.1184	.0049
	r'	.0049	.0707	0
	others	.0239	.0021	0
P	P_1	.5161	.8911	.1677
	P_2	.4839	.1089	.8323
Auberger	Au^a	.6213	.6419	
	Au	.3787	.3581	
Xg	Xg^a	.67	.55	.54
	Xg	.33	.45	.46
Secretor	Se	.5233	.5727	
	se	.4767	.4273	

Source: R. C. Lewontin, *The Genetic Basis of Evolutionary Change* (Columbia University Press, 1974).

The three polymorphic genes for which the major geographical groups are most similar—those for the Auberger, Xg, and Secretor proteins—show how variable a gene can be within populations while having no variation at all between groups. It must be remembered that these are the three most similar *polymorphic* genes. Almost 75% of human genes examined are monomorphic, with no variation either within or between populations. That is, for 75% of the known human genetic endowment, *all* humans are identical irrespective of their geographical origin.

Relative Variation within and between Populations

Data on gene frequencies can be used to answer the question of what fraction of all human diversity is contained within a given population and how much more is added when we consider different groups. There are various measures of the diversity of objects in a collection, all of which are equivalent to asking the probability that two objects taken at random from the collection will be of different kinds. The diversity can be measured at several levels of agglomeration of individuals to see how much is added at each level. Thus, the probability that two people picked at random from a local tribe or nation will be different can be compared with the probability that two people picked at random will be different if several tribes or nations are lumped together in one "race." Further, all of the races can be lumped together to determine how much greater the chance of difference between two randomly chosen people has become. Suppose that there is a lot of variation between individual people within a nation but that all nations and races have exactly the same frequencies of variants within them. Then lumping them together will change nothing, and the diversity among all people in the world will be no greater than the diversity found within a village. On the other hand, suppose that one tribe is 100% type X and another is 100% type Y. There is no diversity at all within each population, but, if the two populations are merged, there is a good chance that two people chosen at random will be different. In such a case, all of the diversity within the species as a whole would be between tribes, and none of it would be within tribes.

Currently, 17 polymorphic human genes have been studied in enough nations and tribes to make possible a calculation of comparative diversity. These genes are listed in the table on the next page, together with the range of allele frequencies known among populations and the names of the populations at the extremes. The most revealing feature of the table is that most of the names are those of small isolated groups of American Indians, South Asian aborigines, linguistic isolates like the Basques, or small island populations. These small, semi-isolated populations lose genetic diversity by chance, and they do not recover it because migration into them from the world at large is so infrequent. Thus, they are likely to have allele frequencies that are very near one extreme or the other of the world distribution. Suppose the frequency of each allele at each known locus were represented along a separate dimension in a multidimensional space. Then a population could be represented as a point in that space whose projection on each of the various axes would be the frequency of each allele. The triangular representation of the ABO blood-group frequencies (p. 117) is a simplified example of such a space. All the human populations in the world would form a cloud of points in the space. In the center of the cloud would be those populations with gene frequencies "typical" of the human species, while around the edges of the cloud would be those populations that were at one extreme or another for various alleles. For the most part, the small, isolated island or aboriginal populations would be far out at the edge of the cloud near the corners of the space, while the large populations of Europe, Asia, and Africa would be clustered together near the center of the mass.

To determine the diversity between "major races," some decisions have to be made about which local populations belong in which races. Once again, we are

Extremes of allele frequency of polymorphic blood groups and proteins in known populations	Locus	Allele	Frequency Range	Extreme Populations
	Haptoglobin	Hp^1	.09–.92	Tamils-Lacondon
	Lipoprotein	Ag^x	.23–.74	Italy-India
	Lipoprotein	Lp^a	.009–.267	Labrador-Germany
	Xm	Xm^a	.260–.335	Easter Is.-U.S. Blacks
	Red cell acid	p^a	.09–.67	Tristan da Cunha-Athabascan
	phosphatase	p^b	.33–.91	Athabascan-Tristan da Cunha
		p^c	0–.08	Many
	6-Phosphogluconate dehydrogenase	PGD^A	.753–1.000	Bhutan-Yucatan
	Phosphoglucomutase	PGM_1	.430–.938	Habbana Jews-Yanomama
	Adenylate kinase	AK^2	0–.130	Africans, Amerinds-Pakistanis
	Kidd	JK^a	.310–1.000	Chinese-Dyaks, Eskimo
	Duffy	Fy^a	.061–1.000	Bantu-Chenchu, Eskimo
	Lewis	Le^b	.298–.667	Lapps-Kapinga
	Kell	K	0–.063	Many-Chenchu
	Lutheran	Lu^a	0–.086	Many-Brazilian Amerinds
	P	P	.179–.838	Chinese-West Africans
	MNS	MS	0–.317	Oceanians-Bloods
		Ms	.192–.747	Papuans-Malays
		NS	0–.213	Borneo, Eskimo-Chenchu
		Ns	.051–.645	Navaho-Palauans
	Rh	CDe	0–.960	Luo-Papuans
		Cde	0–.166	Many-Chenchu
		cDE	0–.308	Luo, Dyak-Japanese
		cdE	0–.174	Many-Ainu
		cDe	0–.865	Many-Luo
		cde	0–.456	Many-Basques
	ABO	I^A	.007–.583	Toba-Bloods
		I^B	0–.297	Amerinds, Austr. Abo.-Toda
		i	.509–.993	Oraon-Toba

Source: R. C. Lewontin, *Evolutionary Biology* 6 (1972): 381–398.

in the old typological bind of the anthropologists: Are Turks Caucausians or Mongoloids? Are Ethiopians Negroid? Where do the Hindi and Urdu speakers go? For the purposes of calculation, eight major races have been delineated: Africans, Amerinds, Australian aborigines, and Oceanians. A few arbitrary decisions were made (Finns and Hungarians were considered Caucasians, but Turks were considered Monogloids), but the results, shown in the table on the next page, turn out not to be very sensitive to some minor reshuffling of local nations into other races. The table shows, in the second column, the amount of genetic diversity contributed to the whole species by each gene. Some genes, such as the adenylate kinase gene, have one allele that is very common every-

Proportion of genetic diversity accounted for within and between populations and races		Proportion		
Gene	Total H_{species}	Within Populations	Within Races between Populations	Between Races
Hp	.994	.893	.051	.056
Ag	.994	.834	—	—
Lp	.639	.939	—	—
Xm	.869	.997	—	—
Ap	.989	.927	.062	.011
6PGD	.327	.875	.058	.067
PGM	.758	.942	.033	.025
Ak	.184	.848	.021	.131
Kidd	.977	.741	.211	.048
Duffy	.938	.636	.105	.259
Lewis	.994	.966	.032	.002
Kell	.189	.901	.073	.026
Lutheran	.153	.694	.214	.092
P	1.000	.949	.029	.022
MNS	1.746	.911	.041	.048
Rh	1.900	.674	.073	.253
ABO	1.241	.907	.063	.030
Mean		.854	.083	.063

Source: R. C. Lewontin, *Genetic Basis of Evolutionary Change* (Columbia University Press, 1974).

where; so they account for very little diversity. Other genes, such as the gene for the Rh blood-group system, vary immensely from person to person and from group to group, giving a larger value of total diversity. The last three columns of the table show what proportion of the total diversity for that gene is (1) between individual people within a local population, (2) between local populations (nations or tribes) within a major racial subdivision, and (3) between major races. Although there is some variation from gene to gene, the result is quite consistent: Of all human genetic variation, 85% is between individual people within a nation or tribe. This figure is not affected, of course, by how the nations are assigned to major races. The remaining variation is split evenly between variation between nations within a race and variation between one major race and another. To put the matter crudely, if, after a great cataclysm, only Africans were left alive, the human species would have retained 93% of its total genetic variation, although the species as a whole would be darker skinned. If the cataclysm were even more extreme and only the Xhosa people of the southern tip of Africa survived, the human species would still retain 80% of its genetic variation. Considered in the context of the evolution of our species, this would be a trivial reduction.

Adaptive Characteristics

The pattern of variation among human groups for molecular polymorphisms shows that there has been rather little genetic differentiation between populations, even between those at opposite ends of the world. Yet this contradicts our observations of the obvious major differences between geographic groups in skin color, stature, body form, hair form, and so forth. There may be no difference between the blood types of a Pygmy and a Swede, but anyone can tell the two apart at a glance. Moreover, overt differences between groups seem not to be random but rather to make some geographical sense. People who live in hot places are dark skinned and dark eyed and dark haired, while northern Europeans are pale. One possibility is that widespread physical similarities represent a historical trace of recent common ancestry. The uniformly black, straight hair both of Mongoloid peoples and of American Indians is probably just the consequence of the origin of the Amerinds from people of Siberia as recently as 12,000 years ago. Other similarities, like the high concentration of brown skin pigment in tropical Africa, southern India, and Australia, cut across lines of ancestry. For these, it is tempting to offer an adaptive explanation, relating the characteristic to the environment in which the population lives.

Some differences between groups are undoubtedly related to environmental exigencies. The high frequency of the gene for hemoglobin S in western and central Africa, in Yemen, and in India is a case in point. The regions of high gene frequency correspond to the regions in which malaria caused by *Plasmodium*

The intensity of skin color varies with latitude but cuts across lines of relationship between groups. The map shows clines in skin-color intensity in reflectance units for populations at different distances north and south of the equator.

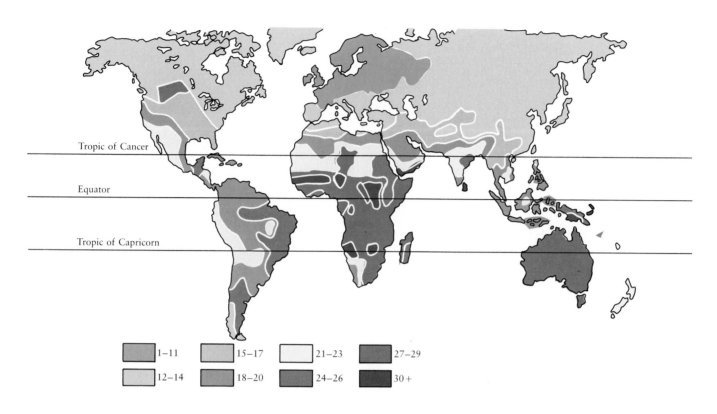

Tropic of Cancer

Equator

Tropic of Capricorn

1–11	15–17	21–23	27–29
12–14	18–20	24–26	30 +

American Indians and Mongoloid peoples have uniformly black, straight hair probably because of their common origins.

falciparum is endemic. Because heterozygosity for the hemoglobin S gene confers resistance to malaria, heterozygotes will survive and leave more offspring in malarial regions. Thus, the gene will have a high frequency there, despite the early death of homozygotes for hemoglobin S. The reasonableness of this explanation is supported by the lower than expected frequency of hemoglobin S in American blacks, who are not routinely exposed to falciparium malaria. A similar explanation (also involving resistance of heterozygotes to malaria) may account for the high frequency of β-thalassemia in the Mediterranean region and of glucose-6-phosphate dehydrogenase enzyme deficiency in the Mediterranean region, central Africa, and India. The map on the facing page shows this frequency relation.

When we turn from molecular diseases to color and shape, the evidence is more problematical. Intensity of brown skin pigment does seem to be correlated with the intensity of ultraviolet radiation. The map on page 125 suggests that the common ancestors of modern Old World populations had an intermediate amount of brown pigment and that, as they spread from Asia into Africa and Europe, the pigment was lost in the north and intensified in the tropical south. The adaptive explanation is somewhat tenuous, however: In theory, skin pigment was intensified in tropical populations as a protection against the harmful effects of intense ultraviolet radiation. This would happen only if people with more skin pigment left more offspring. But even the most devoted Anglo-Saxon sun worshipper, tanning on the beaches of southern California, does not develop skin cancer until middle age. If early hunting and gathering societies were at all like present-day ones, all children would have been produced well before

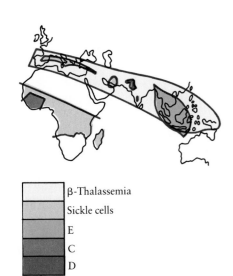

β-Thalassemia
Sickle cells
E
C
D

The distribution in Africa, Asia, and the Mediterranean of various hemoglobin variants and of β-thalassemia.

the age at which skin cancer would develop. Nor can one invoke the loss of support of children from the premature death of parents, inasmuch as the highly cooperative nature of primitive societies protects orphaned children. Among northern Europeans, in contrast with Africans, the problem is an insufficiency of ultraviolet light, which is needed to convert ergosterol into vitamin D in the skin. Although this supposed insufficiency is usually offered as the explanation for the loss of melanin from populations in the north, no one seems ever to have actually tested the rate of vitamin D production in people with different amounts of skin pigment under different amounts of solar irradiation. It would also be necessary to show that differences in vitamin D production would in fact have a differential effect on survival and reproduction. It is yet to be demonstrated that the average Arab, for example, has too much skin pigment to thrive in Helsinki or too little for Dakar.

A somewhat stronger case for adaptive differences between tropical and temperate and arctic people can be made for body shape and metabolic responses to heat and cold. The problem for a mammal is to maintain a constant internal body temperature despite a variation in ambient temperature that may go from below freezing at night to 110°F in the shade during the day in a single 24-hour period. Among human beings, the major responses to large temperature variations are behavioral. People put on or take off clothes, make fires and shelters, go into the shade in the hottest part of the day, huddle together at night, adjust their diets—all to even out the major variations in ambient temperature. Superimposed on the behavioral responses to major temperature variations are physiological and anatomical adjustments that can cope with less severe and shorter stresses. When we are hot, we sweat for evaporative cooling; when we are cold, we erect our body hairs and contract our skin and our peripheral blood vessels in order to conserve heat.

It is reasonable to expect, then, that some differences will have developed among human geographical groups in the way they cope with temperature stress. One method of coping with cold stress is to increase the metabolic rate to increase heat production. Blacks increase their metabolic rate in response to surface chilling more slowly than Europeans, who, in turn, are slower in their response than Eskimos. An alternative method is to reduce metabolic activity, as Australian aborigines do when sleeping, to demand less energy for maintenance. This kind of "hibernation" response appears not to be characteristic of all Australian aborigines, however: Those who live in more temperate coastal regions seem to lack it. There are no generally consistent metabolic differences between groups in response to heat stress, although some studies have shown that Africans, given a standardized work task, sweat less than Europeans under identical conditions.

Physiological responses are not without cost. Keeping warm when it is cold requires burning more reserves of fat and carbohydrate and eating more, but keeping cool is also an energetic process. Many calories are consumed in sweating, and salt is lost. A major problem in tropical agriculture is that cattle consume so many calories in the work of regulating their body temperature that they have little fat left over for milk or meat production. At every age, energy consumption and expenditure are relevant to survival and reproduction. Especially where infectious disease is a powerful source of mortality, the ability to cope with heat stress easily and efficiently may be critical. Thus, differences in body shape become important as an alternative, passive mode of temperature buffering. Because smaller bodies have a greater surface-to-volume ratio than larger ones, they lose heat more rapidly. Elongated bodies have a greater surface-to-volume ratio than more compact ones with shorter arms and legs. Not unexpectedly, on a world-wide basis, there is a negative correlation between body weight and average environmental temperature but a positive correlation between temperature, body length, and limb length. Typically, the Eskimo has a large chunky torso and short limbs, whereas the Dinka of Africa is tall and thin with very long arms and legs. Although these trends seem to make good sense, there is no actual demonstration that they subserve greater survival and reproduction.

Skin color and body shape illustrate the problems of making adaptive claims about human traits. First, exceptions can always be explained away by an appeal to history. The pigmentation of American Indians is more or less the same from Point Barrow through equatorial Brazil to Tierra del Fuego. This uniform-

An Eskimo and a Nilotic Negro have very different ratios of body surface area to body volume.

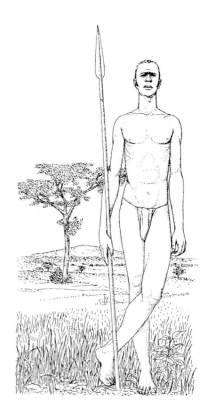

ity is explained as the result of the very recent origin of the Amerinds from their ancestors in Siberia only about 12,000 years ago. Natural selection has not yet had time to effect the accumulation of genes for higher melanin among the Xavante or the Navaho. Second, circumstances are always complex, and there are contradictory pressures. Tropical Africa is the home of both the Pygmies and the Watutsi, among the shortest and tallest people in the world. It can be argued that these are simply two alternative modes of coping with heat stress, being small or being very elongated. Moreover, the Watutsi originated in the dry heat of Ethiopia, where evaporation of sweat is rapid, while the Pygmies live in the moist forest, where perspiration is not much help. By bringing together historical elements and the variability of the environment and of physiology, any difference can be rationalized as adaptive. One must beware of taking such adaptive stories too seriously in the absence of a knowledge of actual data on survivorship and reproduction. Adaptive reconstruction of the causes of human differences remains essentially an amusing pastime, testing our ingenuity as imaginative story-tellers.

Race, Class, and Ability

The inequalities of status, wealth, and power in the world are not distributed at random. There are poor and rich nations and, within nations, social classes to which families belong over many generations. In some countries, like the United States, social class is in part confounded with race and ethnicity. In 1977, the median annual income of white American families was $16,740, but that of blacks was only $9,563, a discrepancy that had increased over the preceding ten years. The table below shows other dimensions of the long-term inequality by race in America. The existence of permanent hereditary underclasses is even more of a challenge to political notions of freedom and equality than is inequality among individual people. Temporary differences in fortune during a person's lifetime can be laid to bad luck, either of circumstances or of parentage. But when whole groups, like blacks, continue generation after generation to have higher unemployment rates, lower incomes, fewer years of schooling, and greater mortality rates, something more than bad luck must be involved. An explanation that has long been offered for inequalities of race and class is an extension of the biologist's explanation of individual differences. In outline, the argument runs as follows:

1. There are major differences between races and social classes in IQ performance.
2. IQ measures intellectual ability.
3. Social success depends upon intellectual ability.
4. IQ is highly heritable.
5. Highly heritable traits are not susceptible to change by social arrangements.

Therefore:

6. Differences between races and classes are probably also genetic.
7. Differences between races and classes are fixed and not susceptible to change by social arrangements.

Indexes of economic and social status of American blacks and whites	1950		1970	
	Whites	Blacks	Whites	Blacks
Median family income	$3,445	$1,869	$10,236	$6,516
Percent completed high school	33.6	12.1	57.2	35.4
Percent managers and technicians	17.0	3.8	22.5	12.6
Percent unemployed	4.9	9.0	4.5	8.2
Infant and fetal mortality (per 1,000 births)	63.3	104.5	44.0	76.9
Life expectancy (males)	66.5	59.1	68.0	61.3

Source: Data from U.S. Bureau of the Census.

Mean IQs of preschool children classified by fathers' occupation		
Father's Occupation	IQ	IQ
Professional	116	116
Semiprofessional and managerial	112	112
Clerical and skilled trades	108	108
Semiskilled and minor clerical	105	104
Slightly skilled	104	95
Unskilled	96	94

Source: Data provided by M. Schiff.

How well does this argument hold together? There are indeed major differences in IQ performance between classes and racial groups. The table at the left shows the mean IQ performances of preschool children according to their fathers' occupations. The data are from two different studies. The most extensive report of the relationship between IQ and parents' occupation is, unfortunately, the work of Cyril Burt. Not unexpectedly, he reported a discrepancy between social classes even greater than that shown in the table. The discrepancy between blacks and whites in the general school population is equally clear. The white school population, on whom the tests are standardized, has a mean IQ of 100 and a standard deviation of 15. The mean of blacks is 85, about one standard deviation below that of the whites. So the first assertion of the argument appears to be quite correct. All of the rest, however, are either factually incorrect or conceptually false.

In the first place, we have already seen (pp. 94–95) that whatever it is that is required for social success, IQ tests do not measure it. The high correlation between IQ and social status is the result of indirect causal paths between parents' social status, performance on IQ tests, amounts of schooling, and eventual social position. For persons of equal IQ, family background is an excellent prediction of eventual success, while for persons of equal family background, IQ has virtually no predictive power (see the graphs on p. 96).

Second, we have seen that the evidence for the high biological heritability of IQ scores is, at best, dubious because it confounds biological similarity with environmental similarity. More important, the heritability of a trait is not an index of its susceptibility to environmental and social change. In adoption studies, children adopted by middle-class families typically have higher IQs than their biological parents or than children who are left in orphanages or are returned to their own families. In Skodak and Skeels' study, for example (pp. 102–103), the adopted children's mean IQ was 117, whereas their biological mothers' mean IQ was only 86. If there is one point that is made over and over again in this book, it is that heritability does not mean fixity or even resistance to change.

Third, there is a confusion between the causes of variation within populations and causes of variation between populations. The fact that a trait has very high heritability within a population provides no evidence about whether the difference between populations has a genetic basis. The causes of variation among individual people are not the same historically or physically as the causes of differences between groups. This can be illustrated by a hypothetical but biologically realistic example. Suppose I have two highly inbred strains of corn, each of which is totally uniform genetically. The two strains differ, however, because they are homozygous for different alleles of various genes. Suppose, further,

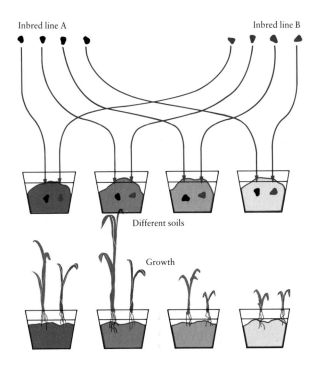

Different soils

Growth

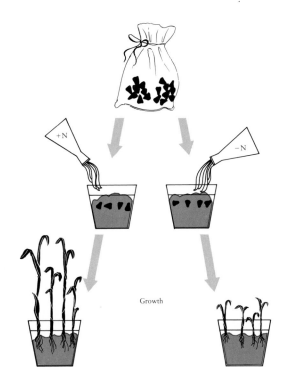

Growth

that I plant one seed from each of the two strains in each of 100 pots of soil taken from a variety of different regions of the country. When the seeds produce plants, I will find that, for each strain, there is variation from pot to pot in plant height because some pots have good soil and some have poor soil. This variation within each strain has zero heritability because there is no genetic variation at all within each strain. All of the pot-to-pot variation is environmental. When I compare the two strains, however, I discover that strain *A* has grown better than strain *B* in every pot. That difference between *A* and *B* is a genetic one, because the two strains experienced identical sets of environments. Thus, we have a difference *between* two strains that is entirely *genetic*, even though the heritability of the trait *within* the two strains is zero.

Consider, now, a different experiment. Suppose I take two handfuls of seed from a sack of an open-pollinated variety of corn. There is a lot of genetic variation from seed to seed in such a variety, but there will be no average genetic difference between the seeds in my left hand and those in my right. The seeds from one hand are planted in washed sand to which a carefully compounded plant-growth solution is added. The seeds from the other hand are treated in exactly the same way, except half the nitrogen is left out of the solution. When the plants have grown, there will be variation in height from plant to plant within each lot. That variation will be totally genetic: All of the seeds in each lot were grown under identical conditions, but there were genetic differences between seeds. Plant height has 100% heritability in this case. There will also be a big difference in average height between the lot grown with the complete plant-growth solution and the lot grown with a solution from which half of the nitrogen was omitted, but that difference will have no genetic component. It will be a consequence of the lack of nitrogen. Thus, we have a difference *between* populations that is totally environmental, even though the heritability of the trait *within* populations is 100%.

If one wishes to make an inference about the genetic differences between groups, only observations about the groups as whole entities are relevant. Nothing can be learned simply from examining the genetic differences among individuals within groups. For a trait like IQ performance, what we need is information about how blacks and whites would perform on IQ tests if they had the same family histories, or if somehow differences in family history were independent of ancestry. It is hard to imagine such circumstances, especially in societies where race is of such immense social importance, but a handful of observations do exist. They are very revealing.

Tizard's study of English orphanage children included those with white, black, and mixed parentage. Children were taken into residential nurseries, most of them being admitted before they were a year old. The children were given nonverbal intelligence tests at an early age (from 2 to 5 years old), but only after they had been in the orphanages for at least six months. The results are given in the table at the left. There are indeed consistent differences in IQ performance—differences in favor of blacks—though only on Test 3 are there differences greater than those expected from chance experimental variation.

A second example is provided by the children of American soldiers of occupation who were left behind to be raised by their German mothers when their fathers returned home. When tested in the German schools, the children of black fathers did slightly, but not significantly, better than the children of white fathers.

These studies, and two that show no correlation between the IQs of black children and the probable amount of their white ancestry, provide all the evidence we have about IQ performance when environments are randomized with respect to racial ancestry. They all point to the same conclusion: There are no biological differences between blacks and whites in whatever it is that is measured by IQ tests.

Scores of residential nursery children on three nonverbal IQ tests

Children of	Test 1	Test 2	Test 3
White parents	102.6	98.5	101.3
Black parents	106.9	97.8	105.7
Mixed parents	105.7	99.3	109.8

Source: B. Tizard, *Nature* 247 (1974): 316.

Aside from IQ, there is a great deal of folklore about differential abilities and temperaments among ethnic racial groups and social classes. In fact, however, there is no evidence that any of these supposed differences is anything but culturally and historically contingent. Again, musical ability provides a kind of paradigm: At various times, various nationalities have been thought to have inherently superior musical talents. In Mozart's time, it was the Italians; in the nineteenth century, it was the Germans. Anyone brought up on classical music during the middle third of the present century "knows" that Eastern Europeans, especially Jews, have a special talent for string playing. Yet, in the past 10 years, we have seen a progressive replacement of people with names like Jascha Heifetz and Emmanuel Feuermann by others with names like Young-Uck Kim and Yo-Yo Ma. The gene for instrumental playing seems to have migrated from Jews to Asians in a remarkably short time.

Male and Female

Cutting across the differences between races, ethnic groups, and social classes is that distinction that is primary in our social consciousness and in our self-images: the difference between male and female. The anatomical distinction—which, with few exceptions, is unambiguous at birth—becomes greater from childhood to adulthood, and, at the same time, a distinction in self-image, in

overt behavior, and in social roles develops in parallel with the developing anatomy. Every society of which we have a record has some division of labor by sex, although the particular tasks that are regarded as "men's work" or "women's work" vary considerably. There are men's rites and women's rites, men's songs and women's songs, men's fashion and women's fashion, things that are forbidden to men and things that are forbidden to women. Every society in every era has used the anatomical and reproductive dichotomy between male and female as the basis for a deep dichotomization of social organization along both productive and ritual lines. The ubiquity of this cultural distinction between the male and the female has often been offered as evidence that social differences between the sexes are in the genes, that "anatomy is destiny." After all, boys and girls really are different from birth, as any fool can see. It seems almost self-evident that differences in behavior and social role between the sexes are part of that same biological differentiation that enables the doctor to answer, in response to the first question parents ask about the newborn baby, "It's a boy." The truth, however, is more complex and more subtle.

It is commonplace to say that the difference between boys and girls is genetic. Yet sex has no heritability, in the sense that height does. Tall parents may have tall children, and short parents may have short children, but every boy and

The human chromosome set in a male. A female would have two X chromosomes of identical shape, instead of an X and a small Y.

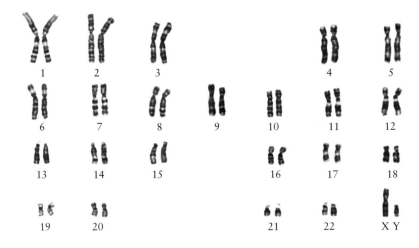

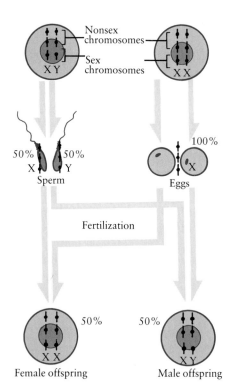

Why the segregation of chromosomes in gamete formation leads to a 50:50 ratio of the two sexes.

every girl have exactly one parent of each sex. Nor has anyone ever discovered a gene like the blood group of enzyme-coding genes that is present in one sex but absent in the other.

The difference between the sexes is genetic in a rather odd sense. Humans have twenty-three pairs of chromosomes, one member of each pair having been contributed by the sperm and one by the egg. For twenty-two of these pairs, the two members of each pair are identical with each other in form. In females, the twenty-third pair is also made of two identical members. In males, however, the two chromosomes of the twenty-third pair are quite different from each other. One member, the X chromosome, is identical in form with the pair in the female, but the other member, the Y chromosome, is much smaller (see the figure above). Thus, females have two X chromosomes, while males have one X and one Y.

The chromosomal difference between males and females explains the way in which sex is "inherited" and why the ratio of boys to girls is close to 1:1. When gametes are formed, one member of each chromosome pair is distributed to each gamete. Because females have the chromosomal constitution XX, every egg cell will contain one X chromosome. Males, however, are XY, and so half of all sperm cells will carry an X chromosome and half will carry a Y. When an X-bearing sperm fertilizes an egg, the consequence will be an XX zygote, a female. A Y-bearing sperm will fertilize an egg to produce an XY zygote, a male. A female gets one of her X chromosomes from her mother and one from her father, while a male gets his X chromosomes from his mother and his Y chromosome from his father (see the diagram at the left). We expect a 1:1 sex ratio because X-bearing and Y-bearing sperm cells should be equally common. However, there are small deviations from this expectation at birth, as a result of slight differences in the ability of X-bearing and Y-bearing sperm to fertilize eggs and differences in prenatal mortality of male and female fetuses. The latter is probably quite important, as shown by historical changes in the sex ratio at birth. In the 1840s in Great Britain, there were about 105 boys born alive per 100 girls; this decreased to about 104:100 in 1900 and rose to more than 106:100 by the 1960s.

The presence of one X and one Y in males and two X chromosomes in females suggests that males will possess genes, on the Y chromosome, that females do not, while females will have twice as many genes as males for the genetic factors on the X chromosome. But life confounds logic. Although about 90 genes are now known to occur on the X chromosome—including those for red-green colorblindness, the Xg blood group, the enzyme glucose-6-phosphate dehydrogenase, and hemoglobin—no genes have ever been located for certain in the Y chromosome. Thus, there are no known "male only" genes. Conversely, since males have an X chromosome, there are no genes that are carried by females only. The lack of known genes does not mean that the Y chromosome has no developmental effect. In about 1 in 4000 births, a person is born who has one X chromosome but no Y. If the Y chromosome were totally without effect in development, that XO person would be male, and we would conclude that it is the number of X chromosomes that determines whether a person is male or female. In fact, XO people are *females*, although their ovaries are poorly developed and they are infertile. They also have a number of growth abnormalities, which, taken together, are called Turner's syndrome. We must conclude that the Y chromosome does, in fact, push the development of the embryo toward maleness. This influence is confirmed by the rare occurrence of people who have two X chromosomes and a Y chromosome. They are unmistakably male, but they are sterile and suffer from an assortment of developmental defects called Klinefelter's syndrome.

The presence of two X chromosomes in females does not mean that they have twice as many effective X-chromosome genes as males. Early in the development of a female, one of the X chromosomes in most cells becomes inactive, and the genes it carries are switched off. In some cells, it is the paternally inherited X chromosome that is inactivated; in others, it is the maternally inherited one. Apparently, this inactivation occurs at random; but once it has taken place in a given cell, all cells that arise from it in future development will have the same chromosome inactive. The inactivation of genes is accompanied by a change in the form of the chromosome itself. It becomes what is referred to as a Barr body, a highly condensed lump that stains much more darkly with chromosomal dyes than does its active mate. In fact, the presence or absence of Barr bodies in cells

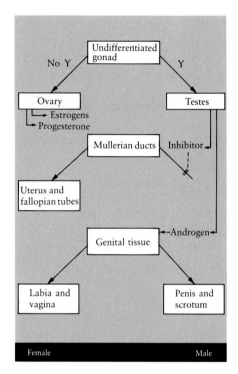

The sequence of events in development of a fetus that leads to male or female primary sexual characteristics under normal circumstances.

taken from a fetus *in utero* can be used to predict the sex of the infant. Despite the inactivation of the genes on one of the X chromosomes, the presence of two X chromosomes has an important effect on development: XO girls have imperfectly developed ovaries, and XXY boys are sterile.

We seem to have a paradox. To be a normally developed, fertile male requires having one X and one Y chromosome, while normal fertile females have two X chromosomes. Yet no genes have been found on the Y chromosome, and one set of genes is turned off in XX females. If males and females have exactly the same assortment of active genes, why are they male and female? The solution to the paradox lies in the relative activities of genes and in the differential reaction of different tissues to the same signals at different moments in development. The gonads of males and females alike secrete the "female" estrogen hormones, but females secrete more than males. The estrogens are not proteins, and so they are not the direct products of genes, but they are synthesized by enzymes that are coded for by genes. Both males and females must have the genes that code for these enzymes, and the genes must be active in both sexes. The difference between the sexes lies, presumably, either in the amount of the enzymes produced or in the cellular conditions for their activity. Moreover, the mere presence of a hormone, no matter what its concentration, is not sufficient to cause sexual differentiation. The tissues must be receptive to the hormone. The presence of the "male" hormone testosterone (also found in lesser amounts in females) usually causes the developing genital tissues of the fetus to produce a penis, a scrotum, and the internal ductwork of a male. Occasionally, however, an XY embryo has genital tissues that are insensitive to the influence of the testosterone produced by its testes; so, instead of developing the external genitalia of a male, it has the outward appearance of a female. Although such cases are rare, they demonstrate that normal sexual differentiation depends both on the nature of the chromosomal and hormonal signals and on the readiness of the developing tissues to respond to those signals. The readiness of the tissues is, in turn, a consequence of previous developmental events. In this way, the original differences in chromosomal composition push the organism into either the male or the female developmental pathway without different genes being involved at every step of the way.

Our current understanding of how the successive steps in sexual development occur up to the moment of birth is shown in the diagram at the left. Embryos begin with an undifferentiated pair of gonads. The presence of a Y chromosome induces those gonads to become testes. Otherwise, they develop into ovaries. Once this change has occurred, the rest of development unfolds apparently without any further chromosomal signals. The testes produce androgen hormones, which cause the genital tissues to develop into the external genitals of the male. Without that stimulus, those same tissues would develop into female genitals. The testes also produce an inhibitor that prevents the rudimentary internal ductwork from becoming a uterus and Fallopian tubes, as it would in the course of female development. In summary, an early human embryo has the possibility of following either one of two alternative developmental tracks. The presence or absence of a Y chromosome acts as a switch, early in development, to shunt the organism into one of the alternative pathways. The rest is history.

Man and Woman

The anatomical and hormonal differences between males and females are only a small part of the difference between the sexes. The social reality of sex is that there are differences in self-image, in power, in social roles, in the distribution of males and females by occupations, by personality types, and by causes of sickness and death. What is the origin of these differences in what psychologists J. Money and A. A. Ehrhardt have called *gender,* as distinct from merely physical sex? How do the social phenomena of man and woman arise from the biological phenomena of male and female?

There have been two kinds of explanations for gender differences. One, a biologistic explanation, sees gender as being the same as sex. In this view, gender differences arise as the direct consequences of the same chromosomal and hormonal causes that produce gonadal differentiation. The X and Y chromosomes and the hormones are said to influence the development of the nervous system so that the brains of female babies differ systematically from the brains of male babies. These differences in brain structure are then thought to be reflected in differences in abilities, attitudes, and desires. In this view, hormonal differences in later life also affect the behavior of men and women directly, making men more aggressive and women more capricious.

The other view is that gender differences are not the direct consequences of chromosomal and hormonal differences but that they arise from the socialization and self-images of people, identified from birth by their external appearances, as male or female. That is, gender differences are based on, but are not caused by, sex differences. Parents identify a baby as female, socialize it as a female, and make it aware, at an early age, of its own female identity. In this way, individual people and society interact to determine each person's gender identity. The diagram on the next page shows one version of this postulated chain of circumstances from the chromosome to gender. It includes, as a broken line, the possible influences of hormones on brain and of brain on gender, but these are not regarded as being of primary importance.

There is, in fact, a great deal of evidence on the question of whether chromosomal and hormonal differences directly affect gender. This evidence comes from a number of "natural experiments" and from deliberate manipulations of

The paths of causation that lead a person with male or female primary sexual traits to take on a masculine or a feminine gender.

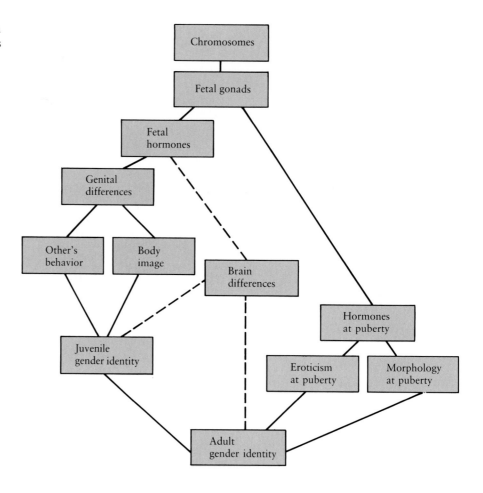

various aspects of people's sex. There are, for example, people with *testicular feminization syndrome*. These people have a male (XY) chromosome constitution, testes, and normal male hormone proportions, but because their genital tissue was insensitive to testosterone during development, they grew up with female external genitalia, including a vagina. Such people are raised as girls, and their abnormality is discovered only when, as young women they fail to menstruate. They provide a test of the role of chromosomes and hormones as opposed to socialization and self-image. A complementary case is *androgenital syndrome,* in which XX females with normal ovaries develop male external genitalia because of the secretion of extra androgen hormones by their adrenal glands before birth. Some AGS children have ambiguous genitalia, which may be surgically altered to make them unambiguously female. The sum total of

many observations of such abnormal cases is clear in its main result: Children who are seen by others as male, who are treated by others as male, and who therefore see themselves as male develop clear and unambiguous male gender identities. Similarly, children who are assigned to the female sex develop female gender identities. These gender identities are independent of their actual chromosomal or internal morphological or hormonal constitution. For example, Money and Ehrhardt reported on two XX people with androgenital syndrome who had ambiguous external genitalia. One was raised as a girl after her chromosomal constitution was established at age 2 months. The other was raised as a boy. When seen as teenagers, each had the normal gender identity for the sex in which he or she had been raised. An extraordinary and rather grim case makes the same point even more dramatically: One of a pair of identical male twins suffered serious damage to his penis at circumcision. It was decided to reconstruct the child surgically as a female and to raise it as a girl. At the age of 6 years, these identical twins showed all of the stereotyped behavior of a little boy and little girl, respectively. The boy liked to play with toy cars and trucks, was dirty most of the time, and wanted to be a policeman or a fireman when he grew up. The girl dressed up, liked to play with dolls, wanted to get married when she grew up, and helped with the housework.

One must be somewhat cautious about the outcomes of such extraordinary cases. Just because they are so extraordinary, they may not reveal the full subtlety of normal development. Parents of children to whom a gender is assigned by fiat may be more extreme in the degree to which they polarize the children's gender images. Exceptional cases are exceptional. At the same time, there is some evidence, not very compelling, that there are subtle differences in gender identity between people with the normal array of sexual traits and those who are discordant in one way or another. For example, AGS children raised as girls and whose external genitals had been surgically corrected nevertheless were more often reported to be "tomboys," to be less interested in marriage, and to be more interested in careers and sports than girls with normal prenatal hormone balance. There are conflicting reports about the degree of masculinization or feminization of attitudes of children whose mothers were treated with large doses of hormones during pregnancy. None of the results are very clear, there are problems of matching the experimental children with suitable control children of the same background, and many such children have had rather traumatic childhoods. What is clear, however, is that, if there are some effects of hormones on attitudes, these effects are small and well within the normal range of behavior for boys or girls. The primary self-identification of a person as a man or a woman, with the multitude of attitudes, ideas, and desires that accompany that identification, depends on what label was attached to him or her as a child. In the normal course of events, these labels correspond to a consistent biological difference in chromosomes, hormones, and morphology. Thus, biological differences become a signal for, rather than a cause of, differentiation in social roles.

We must treat with extreme caution claims that one or another "masculine" or "feminine" social trait is biologically determined, if for no other reason than the all-encompassing sweep of these claims. Everything associated with gender, from cultural universals like the nurturing of infants by their mothers to the most specific historically and culturally bound activities, such as different trades and crafts for the two sexes in modern industrial society, have been said to be biological. In some cases, occupations are thought to be essentially female, but, when men do enter them, they are said to bring a higher creativity to them. Women are naturally cooks, "but all the great chefs are men." Some of these claims we know to be false simply from historical evidence. Knitting and hand weaving, almost exclusively women's work now that they are outside the mainstream of production, were entirely men's work as recently as 150 years ago. Arguments that physical size and strength make some activities masculine are no longer convincing in the face of historical change. There are now hundreds of women coal miners in the United States; and I attended, the day before writing these lines, the most prestigious chamber-music festival in America, at which women virtuosi performed on such "masculine" instruments as the tympani, the double bass, and the French horn. Claims that certain creative cognitive skills are rarer in women than in men for biological reasons seem to confuse observations and their explanations. Thus, a great deal of publicity has recently been given to a study showing that girls' performance on tests of mathematical ability are, on the average, about one standard deviation below that of boys. The authors of the study come down on the side of genes and hormones causing differences in brain structure because, according to them, the difference between the sexes is too large to be accounted for in any other way. We are not told what priniciples of genetics or social theory connect the size of a difference with its cause.

The explanation of gender differences as a consequence of socialization leaves open for speculation the question of whether the actual content of role differences corresponds historically, or even evolutionarily, to actual physical differences. Because women are the actual bearers of children, and because, on the average, they are somewhat shorter and slighter than men, there may have been in the past a division of productive and reproductive labor that has manifested itself in all human societies. Indeed, no historical transformation can change the division of reproductive labor in its most basic form. But the historical origin of the social differentiation of roles is purely a matter of speculation. What is certain is that the division of productive labor no longer corresponds to average biological differences in stature and strength. What is equally certain is that the immense superstructure of attitude and social power that has been built historically on the base of biological differences has long ago become independent of the actuality of that biology.

Variation in Brazil

There are few countries in the world with a more diverse population than Brazil. The Portugese began to colonize Brazil in the sixteenth century, in the south around São Paulo and in the northeast, and they were soon followed by French Huguenot and Dutch settlers. Almost immediately, children of mixed European and Indian ancestry appeared. These *Mamelucos* (so called for their fancied resemblance to the mixed Turkish-Arab rulers and soldiers of the time) joined exploring bands of Portugese who penetrated into all regions of Brazil, continuing the interbreeding between Europeans and Indians. At the same time, there were massive introductions of African slaves, beginning in the sixteenth century and not ending until 1850. The abolition of slavery at the end of the nineteenth century and the development of gold mining and railroads created a labor market for Asians who came in large numbers from China and Japan. There were further waves of European settlers from Germany and Italy in both the nineteenth and the twentieth centuries, and a certain number of North Americans from the southern United States emigrated to Brazil after the destruction of the plantation economy by the Civil War. Names of Brazilian presidents and generals, such as Kubitschek, Medici, and Geisel, testify to the diversity of European origin of the Brazilian élite. Despite the extensive mixing of Indian, African, Asian, and European ancestry, that political and social élite consists almost entirely of *brancos* (whites) who occasionally boast of a remote Indian ancestor to prove the early arrival of their ancestors. Peasants and lower classes, as these photographs show, display much more of the rich diversity of Brazilian origins.

The Evolution of Human Diversity

9

The central change in the view of the world that took place in the nineteenth century was the rise of evolutionism. Long before the appearance of Charles Darwin's *Origin of Species* in 1859, European thought was becoming committed to the view that change was characteristic of all social and natural systems. In 1795, James Hutton enunciated his theory of the evolution of geological formations, and, in 1796, Pierre Simon, Marquis de Laplace, presented his nebular hypothesis for the origin of the solar system. In the same year, Charles Darwin's grandfather, Erasmus, published a theory of the evolution of life in the form of an epic poem, *Zoonomia.* In 1824, Nicolas Sadi Carnot founded the evolutionary theory of thermodynamics with his demonstration that entropy always increased because all devices that did thermal work were less than perfectly efficient. By the time of the appearance of the *Origin of Species,* Herbert Spencer could support it by arguing that organic life must have evolved since, after all, everything else had.

A world brought into being by means of special creation has no past and no future because the past and the future are exactly like the present: Species began as they now are, and they will remain so, world without end. It is not simply that their history is irrelevant; they are without history. An evolutionary world view, a view in which all systems are in a constant state of flux, differentiates past, present, and future. Adopting such a view is the first step in making descriptions of nature historical. But it is only the first step. To turn mere chronicles of past events into historical explanations, we require that the changes in a system be related to each other in a causal fashion. The present must not simply follow the past, it must be a consequence of it. "And God said, Let there be light: and there was light." This is the language of chronicles, not of history (although, of course, it is implied that there was light *because* God said so). The simple observation that life has evolved—that forms that existed in the past no longer exist, whereas those that live today were absent millions of years ago—is not the same as a theory of evolution. Fossils are a chronicle of past life; they are not a history of past events. Such a history demands a causal theory of how and why one form became another. Darwin's theory of evolution by means of natural selection provided just such a causal explanation that converted a chronicle into a history. It is for that reason that we, quite correctly, associate Darwin's name with the science of evolution, although his grandfather knew as well as he that life had evolved.

A causal theory of change, like Darwin's theory of natural selection, has two implications for the past and the future. Because the present flows continuously and causally out of the past, we can understand our present state fully only by knowing where we came from. The past is the initial condition for the dynamic process that gave rise to the present. It is sometimes thought that it is sufficient

A VENERABLE ORANG-OUTANG.
A CONTRIBUTION TO UNNATURAL HISTORY.

A not entirely sympathetic view of Darwin by a contemporary.

to know the forces operating in a system in order to predict where it will go in its evolution, but that is wrong. If I want to know why I have arrived in Marlboro, Vermont, it is not sufficient to know that I have traveled west at 20 miles per hour for about half and hour. I must also know that I began in Brattleboro. This is the sense in which the past is part of the explanation of the present. It helps us to know why, out of many possible outcomes of the same evolutionary forces, we have arrived at our present form rather than at some other. On the other hand, we must not make the mistake of supposing that the past alone is a sufficient explanation of the present. Human social relations cannot be explained simply by saying that we evolved from apelike ancestors, that we are nothing but "naked apes," all of whose relations are prefigured in the behavior of chimpanzees. Such a view destroys history, in the guise of explaining it, by claiming that our present state is nothing but the past with a different superficial appearance.

The second consequence of a causal theory of evolution is that the future cannot be predicted without reference to the present. Thus, just as the past was the initial condition for the present, what we are now contains what we may become. No matter how useful it might be for an organism to be able to fly, no vertebrate animal has ever been able to develop wings except at the expense of one pair of limbs. It appears not to be within the evolutionary possibilities of tetrapods to develop an extra set of limbs, no matter how convenient that outcome might be. Pegasus was a developmental impossibility. If we are to make any prediction at all, no matter how weak or how optimistic, about the biological future of the human species, we must understand its current biological state.

Because the past is a condition of the present and the present is a condition of the future, it is tempting to assert that we cannot predict what is to come without a knowledge of what has been. In general, however, that assertion is wrong. Although it may seem paradoxical, the relevance of the past for the present does not carry over into the future for most physical systems. That is, what happens next depends only on the present state of the system, not on how it arrived at that state. If I leave Marlboro and travel south for a half hour, I shall arrive in West Halifax irrespective of whether I originally got to Marlboro from the north, south, east, or west. The dynamics of travel does not depend upon a memory of the past. Such systems, in which the future depends on the present but not on how the present was arrived at, are called *Markovian* processes, after the mathematician who first studied them. Most population processes have this Markovian property, as does any physical process that cannot store up information about past events. Thus, the size of the American population in 1982 depends only on how many people of different ages were alive in 1981 and on the birth and death rates among people of the various age classes in that year.

It makes no difference at all whether the population was larger or smaller in 1980 than it was in 1981. Not all physical systems are Markovian. For example, the next word that I write on this page depends not only on the preceding word, but on every other word that I have written. Moreover, since the next word also depends on my intentions, it may very well depend on every word that I have *ever* written (or read). And, as the book grows longer, every word in it will depend on a longer and longer sequence of preceding words.

The distinction between Markovian and non-Markovian processes is fundamental to the difference between human cultural history and human biological evolution. Language, writing, cultural artifacts (such as buildings), and cultural phenomena (such as modes of production) all provide direct information about the past that influences the future. The low state of European culture that continued long after the disintegration of Rome was at least in part a consequence of the immense loss from the fund of technical and humanistic knowledge that occurred at the final destruction of the libraries of Alexandria in 391. In contrast, Muslim culture grew at a prodigious rate beginning in the seventh century partly because the knowledge of classical times—knowledge then unavailable to the Latin and Greek scholars of Europe—was preserved in Arabic manuscripts. The history of a species' biological evolution, however, is stored nowhere in the individual members of the species. Their present state is, indeed, a *consequence* of their history, but the genes currently possessed by the species are all that matter for its evolutionary future, irrespective of how it acquired those genes. There is no "race memory" in biology, only in books.

Darwinian Evolution

Natural systems that change over time, that evolve, do so by two very different mechanisms. Some, like the stars, undergo *transformational* evolution; others, like living organisms, evolve by a *variational* process. Transformational changes are those that occur because every individual member of the system itself undergoes the same succession of stages. The collection changes because all the individuals are themselves developing. The collection of children born in 1960 became a population of adults in 1981 because each child followed a developmental program that turns an individual infant into an individual adult. Stars evolve by burning out, by changing from main-sequence stars, to red giants, to white dwarfs, and then to dead masses. The entire universe is evolving because every star in it—our own sun included—is undergoing transformation over time.

Variational evolution, in contrast, is a process in which the proportions of different kinds of objects in a system change even though the objects themselves do not change. In the United States, about 104 males are born for every 100 females, but, by age 60, there are only about 89 males for every 100 females.

That is not because males have turned into females, like stars turning into red giants, but because males have only about 85% as much chance of surviving to age 60 as females do. The differential survival rate of the two sexes results in an enrichment of the population in the proportion of females.

Before Darwin's *Origin of Species,* theories of organic evolution were transformational. They assumed that the individual members of a species would undergo some change to turn them and their offspring into the members of another species. This was Lamarck's theory, that character changes acquired during an organism's life would be transmitted to its offspring. In this way, the entire species would change in accordance with physical alterations induced in each individual by its interaction with the environment. Giraffes would stretch their necks in order to reach the leaves of tall trees, and their offspring would have longer necks as a result. Darwin's solution to the problem of the origin of new species was radically different. He observed, within each species, a great deal of variation from one individual organism to another. In Darwin's theory the frequency of these variations in the population increased or decreased because of the differential survival and reproduction of the already existing variant forms. The source of variation was, for Darwin, quite a different matter than the process of enrichment of the population in the proportion of one type as opposed to another. The Darwinian theory of the mechanism of evolution is that variation between individual organisms within a population is converted into variation between populations and hence, to variation between species. The difference between species is already immanent in the variation between individual organisms.

The theory of evolution by means of natural selection consists of three fundamental propositions and a mechanical explanation. The three propositions are:

1. Within a population, there is variation among individuals in shape, size, physiology, and behavior (the principle of *variation*).

2. There is a correlation between parents and their offspring such that offspring resemble their parents more than they resemble unrelated individuals (the principle of *heredity*).

3. Some variant forms survive and leave offspring more frequently than other forms (the principle of *selection*).

The mechanical explanation is:

4. The reason some forms leave more offspring than others is that resources are in short supply and some forms are better than others at obtaining them (the principle of the *struggle for existence*).

The cornerstone of evolution by natural selection is the principle of variation. There must be something to "select." If all the individual members of a population are identical with respect to a given character, then, irrespective of how good or how bad that character may be for the species, no evolution will occur. If everyone were equally thin, it would be quite pointless to say that the species ought to develop body fat to protect it from the cold, because there are no fat individuals to select. The principle of heredity is equally essential. Even if there is variation among individual members of a population, differential reproduction cannot change the composition of the population if the offspring of the different types are not different from each other in the same way in which their parents are different from each other. If there were fat people and thin people, and if fat people left more offspring, nothing would change unless the children of fat people were themselves fatter than the average. Evolution by natural selection requires not only variation but also *heritable* variation. Finally, of course, evolution by natural selection works only if different heritable types leave different numbers of offspring. This is the principle of selection. We must not be led astray, however, by the popular characterization of selection as "the survival of the fittest." The word "fit" has many everyday connotations—physically fit, morally fit, and so forth—but none of these is what the evolutionist means by fitness. All that matters for evolutionary change is survival and reproduction. In evolutionary terms, an Olympic athlete who never has any children has a fitness of zero whereas J.S. Bach, who was sedentary and very much overweight, had an unusually high Darwinian fitness by virtue of his having been the father of twenty children.

It is important to realize that the "struggle for existence" is not a necessary part of the variational theory of evolution. Darwin believed in the hypothesis that all organisms produce more offspring than can be supported by the existing resources, a notion made popular by the Rev. Thomas Robert Malthus in *An Essay on the Principle of Population* (which was first published around the end of the eighteenth century). But competition for resources in short supply is not at all the rule among organisms, nor is it the sole cause of differential survival and reproduction. For example, a type of beetle whose larvae can survive freezing temperatures better than those of another type will increase in frequency in the population, if there is a series of severe winters, although it is in no sense competing with other members of its population for some limited resource. Resistance to DDT is now the rule in house flies because the resistant flies survived and left offspring. By no means does evolution depend upon crowding and competition.

Evolution by means of natural selection is self-terminating. The selection process cannot operate unless there is heritable variation to select; but selection means enriching the population for one of the variants, that with the highest fitness, often until the population consists only of that variant. Thus, selection can cause a population to lose the genetic variation it started with and become homogeneous. Selection is like a fire that consumes its own fuel. Once that fuel, variation, is consumed, the process of selection must come to a stop, and evolution ceases. This paradox of variational evolution, that natural selection destroys the very condition that makes it possible, is a critical one: Unless variation is somehow renewed periodically, evolution would have come to a stop almost at its inception.

Sources of Variation

Ultimately, all genetic variation arises from mutation. Errors occur in DNA replication (see Chapter 4), with the consequence that a new amino acid may be substituted in the protein for which the gene codes. Such an altered protein may then have a slightly different sensitivity to temperature or pH, or a slightly altered affinity for the compound on which it acts. But all of evolution cannot consist of changes in genes that are already present. Totally new functions have also been added to the biochemical repertoires of organisms during evolution. Bacteria do not make muscle fibers, bones, hormones, blood antigens, or the host of other structures and compounds that make up higher organisms. New genes, in addition to the old ones, must be made. It is thought that this occurs as a two-step process. First, a gene (or group of genes) is accidentally duplicated so that a chromosome now carries an extra copy of it. Because only one good copy is needed to produce the original protein, the extra copy is free to accumulate mutations without harming the organism. After a time, enough changes may have accumulated in the duplicate to give it a new function. Evidence that such duplication and divergence have occurred in evolution can be seen in the similarity of the amino acid sequences of proteins produced by different genes in the same organism. For example, the α, β, γ, and δ chains of human hemoglobin are coded for by four separate genes, but there are great similarities among the proteins. The α chain has 141 amino acids, while the β, γ, and δ chains have 146 each. The β chain is identical with the δ chain at 136 amino acid positions, and the γ chain is identical with the β chain at 107 positions. Even the α chain, which is the most divergent, is identical with the δ chain at 61 amino acid positions. It also appears that the oxygen-carrying protein of muscle, myoglobin, has descended from a duplication of the hemoglobin gene.

Although evolution depends ultimately on the renewal of variation by the occurrence of mutations, the rate of evolution would be very slow if it were driven by mutation alone. Mutations are rare events. It is not easy to measure mutation rates in humans but, based on observations of experimental animals and on the frequencies of rare disorders in children whose parents are known to be free of those traits, the chance that a sperm or an egg carries a newly formed mutation is between 1 in 100,000 and 1 in 1,000,000 for each gene. Different genes have different rates, but most are at the low end of the range. Such low mutation rates mean that evolution driven by mutation alone could occur only very slowly. Suppose that a population consisted entirely of homozygotes for allele A of a gene, but that the mutant allele a appeared as frequently as once in every 100,000 gametes. Then, after one generation, the frequency of allele a in the population would be only 1 in 100,000; in two generations it would be 2 in 100,000; and so on. As the frequency of allele a increases, the rate of change of the population slows down, since there are fewer A alleles to mutate to a alleles; so it will actually take about 70,000 generations to bring the frequency of allele a up to 50% in the population. Because a human generation is about 25 years, this is about 1,750,000 years, a rather long period in human evolution.

The classic science-fiction notion of the human species suddenly developing an entirely new set of traits as a result of mutations after an atomic war is fundamentally flawed. Ionizing radiation (including that produced by nuclear explosions), if it were present in low enough doses that the recipients were able to survive long enough to reproduce, would probably not do more than double the spontaneous mutation rate. The most extreme assumptions lead to a calculation of a mutation rate only ten times the normal rate. Even at such a high rate, only a tiny fraction of the offspring of irradiated people—at most, 1 in 10,000—would carry a mutation for any given gene, and so the species would hardly be transformed overnight. Moreover, unless the irradiation were continued generation after generation, as indeed it would be if the earth were generally polluted by radioactive fallout, the mutation rate would drop back to its old value. In any case, rapid evolution cannot be driven by mutation alone.

Even though the evolution of a particular trait as a result of mutation alone is very slow, the total rate of input of genetic variation in general is quite large. It is not certain how many different genes each of us has. We have enough DNA to make about 3 million genes the size of those that code for the hemoglobin chains, but a large fraction of the DNA is used up in multiple copies of genes. Even if only 1% of our DNA constitutes distinctly different genes, we must have about 30,000 different genes. If the mutation rate per gene were even as small as 1 in 1,000,000, one new-born child in every thirty would carry a new mutation somewhere in its genes. Thus, new alleles are constantly appearing in members of our species in every generation, although, of course, no one of them is very common.

The variation brought into the human species by mutation is only the base upon which genetic variation is built. The immense richness of gametic types in a species is created, in the end, by sex. Sexual reproduction is a process by which

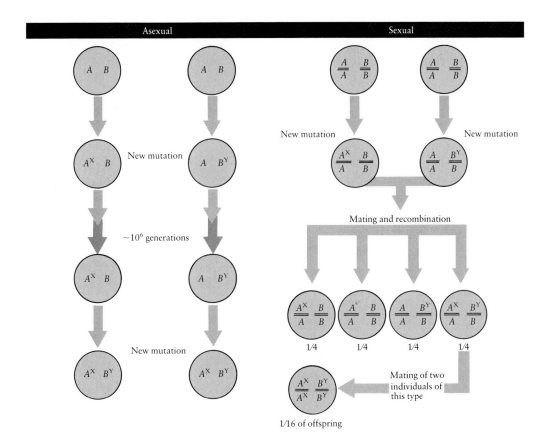

How sexual reproduction leads to the rapid appearance of new combinations of characters as compared with the slow pace of asexual evolution.

the mutational variation among individual people is reconstructed and recombined to produce genetic combinations that were previously rare or absent. This reconstruction occurs in two steps. First, gametes from two totally unrelated family lines are brought together at fertilization, creating a person who now possesses mutations that occurred in quite separate people in the past. Second, when this person later produces gametes, those gametes, as a result of chromosomal recombination, will contain combinations of alleles that previously did not exist in the same gamete. The figure above shows how two mutations arising in separate family lines may in the future be passed along together in a new family line. Suppose that a new mutation at HLA locus A—call it HLA-AX—occurred in one person and that a new mutation at HLA locus B—call it HLA-BY—occurred in another person. In the absence of sexual recombination, there is no way in which these two alleles could ever be present in the same person, except by waiting until, by chance, a person carrying the allele AX also acquires the allele *BY* by mutation. Because mutations are so rare individually, one might have to wait for a very long time. With sexual reproduction, however, the HLA-*AX* and HLA-*BY* mutations may come together at fertilization, and by recombination produce a totally new haplotype, HLA-*AX*, HLA-*BY*. Indeed, the thousands of HLA haplotypes that exist in human populations are certainly the result not of successive mutations occurring in the same family lines but of the recombination of alleles at the four HLA loci, which happens every time an egg or a sperm is formed.

Sexual reproduction makes possible the joining of evolutionary streams from widely separated geographical regions. Conquests, the taking of slaves, the migration of pastoral people, the wholesale immigration of persecuted or impoverished populations into more hospitable societies, all bring people from one population physically into the midst of another. This physical migration is turned into genetic migration by the sexual unions of immigrants and natives, conquerors and conquered, masters and slaves. When the migration is reciprocal (which it rarely has been in human history), both of the original populations will experience an increase in their genetic variation as a result of the infusion of new alleles and new combinations from the migrant individuals. In the more usual, asymmetric, kind of migration, the recipient population becomes enriched genetically to the extent that the donor population differs from it genetically and does, in fact, interbreed with the locals. The population of Australia, at one time almost entirely Anglo-Irish, has become more variable as a result of the immigration of hundreds of thousands of Maltese, Italians, Greeks, and eastern Europeans in the past 30 years. Because differences in allele frequency between races and nations are small for most genes (Chapter 8), migration will not affect genetic variation very much for most characters. For example, the English and Irish have a frequency of the B blood group allele I^B of about 5% whereas southern and eastern Europeans have a frequency of 10%–12%. If 20% of the Australians come from southern and eastern Europe, then the Australian gene frequency for I^B will have risen from 6% to 7%, not a very large change. The main biological effect of migration, in addition to increasing the variation for such characters as skin color, eye color, and hair form, which do differ a great deal between populations, will have been the introduction into the Australian population of genetic variants, such as thalassemia, that were more or less confined to the populations from which the immigrants came. We do not know how many mutations have been added to the Australian population in this way. These biological effects are minor, relative to the immense enrichment of culture that has taken place concurrently. The espresso bar is now a major social institution in Sydney and Melbourne, which once knew only tea rooms and stand-up pubs.

Random Genetic Drift

Every family and every population is limited in size, and each instance of fertilization of an egg by a sperm brings together, by chance, two particular sets of genes. That means that the genetic composition of a population will not be reproduced exactly in each successive generation. Suppose a small village in the Orinoco River drainage consists of 98 type O people and 2 type A heterozygotes $I^A i$. The frequency of the I^A allele is 2/200 = 1% (there are 200 gene copies in 100 people). The two $I^A i$ heterozygotes may, by chance, die without leaving any children, or each may have two children, which is average for village families. In the first case, the frequency of the allele I^A will be different in the next generation than it was among the parents: I^A will be lost entirely to the village population, which will than be totally homozygous *ii*. In the second case, the frequency of I^A might have increased, decreased, or stayed the same: Suppose, by chance, that none of the four children of the type A people were themselves of type A. That is entirely possible because, at each birth, there is a chance of 1/2 that the offspring of a mating of $I^A i$ with *ii* will be *ii*. Under such circumstances, the village population would also have lost the I^A allele. On the other hand, there is also a chance that all four children would have been of type A, in which case the frequency of I^A in the next generation would have risen to 4/200 = 2%. In every generation there will be chance changes in the frequencies of all genes because each generation is, in essence, only a *sample* of the gametes produced by their parents' generation. Once that sampling has taken place and a new generation is established with a new gene frequency, another sampling of gametes will occur in the establishment of the following generation. The process of gene frequency change is Markovian. There is no memory in the present generation of what the original gene frequencies were many generations ago, so that the sampling error accumulates generation after generation. Eventually, by chance, all the copies of one of the alleles will be lost (as in our example of the I^A allele in the Orinocan village) and genetic variation will be lost completely. Until immigration or a new mutation brings the I^A allele back again, the population will remain homogeneous. In probability theory, this process is called the "drunkard's walk," by analogy with a drunkard who emerges from a bar in the middle of the block and takes random staggering steps to the left and right. After each step, it is a matter of sheer chance whether the next step will be in the same or the opposite direction. Eventually, the drunkard must wind up at one end of the block or the other, where he falls into the gutter and goes to sleep. It can be shown mathematically that, given sufficient time, the drunkard is sure to reach one gutter or another—he cannot stagger forever between the ends. Similarly, the frequency of an allele cannot forever oscillate between 0% and 100%; eventually, it must change to produce complete homozygosity. Either the allele is lost totally or it characterizes the entire population and the other alleles are lost.

The consequence of this random genetic drift of alleles in finite populations is that, even in the absence of natural selection, the frequencies of genes change, and, in the long run, genetic variation is lost. It happens quickly in small populations (because only a few gametes are being sampled each generation) and slowly in large populations, but it always happens. Without new mutations, recombination, and migration, every population would eventually become genetically homogeneous and evolution would cease. Again, we can see that mutational change lies at the base of all continued evolution.

Natural Selection

The driving force of evolutionary change is the differential survival and reproduction of the genetic variants that have been produced by mutation, recombination, and migration. The degree of differential fitness associated with various alternative genotypes varies from gene to gene and from circumstance to circumstance. Some genotypes are unconditionally lethal. Homozygotes for the recessive allele that causes Tay-Sachs disease always die because they are deficient in an enzyme that normally breaks down fatty acid alcohols in the central nervous system. There is no environment in which Tay-Sachs children survive. We do not expect such unconditionally lethal alleles to occur in very high frequencies, because natural selection removes them every time a homozygote is born. Presumably, the presence of such alleles in the species is a consequence of repeated mutations that replace them as they are removed by natural selection. Not all lethal alleles are unconditionally lethal, however. Phenylketonuria (PKU) disease provides an example. It is caused by a single mutant allele that, when it occurs in the homozygous state, causes the accumulation of toxic concentrations of phenylalanine, one of the amino acids. People who suffer from this disease are imbeciles or idiots who leave no offspring. The disease can be prevented, however, by restricting the diet so that very little phenylalanine is consumed. The dietary treatment is quite effective, and children treated in this way have normal mental functions, provided that the dietary restriction begins just after birth. Presumably, such treated children eventually reproduce at the same rate as genetically normal people, in which case selection does not operate against the gene.

Tay-Sachs disease and PKU are dramatic examples of natural selection because the effects of genes on survival and fertility are so large. Most natural selection is much weaker and, as a consequence, is quite difficult to measure in human populations. An example is the postulated selection of different alleles at the ABO blood-group locus. There is an association in human populations between being of blood type O and developing a duodenal ulcer, but it is not a very powerful one. Type O people have only 1.4 times the chance of developing a duodenal ulcer as do type A, type B, and type AB people. Moreover, duodenal ulcer is not a major cause of death, accounting for only about 0.4% of the deaths in the U.S. after age 40. In addition, death after 40 has very little effect on reproductive rate. During the prime reproductive period, between ages 20 and 35, only 1% of the total population dies from all causes, and duodenal ulcer accounts for only 0.2% of these deaths, at least in European populations. Taking all ages into account, the selection against type O people from duodenal ulcer is only about 1/10,000 in males and 1/100,000 in females. That is, if 10,000 type A, type B, or type AB fathers were to have a total of 10,000 children, the same number of type O fathers would have only 9,999 children. This very slight difference in reproductive rate would cause only an extremely slow change in the frequency of type O people, and it might easily be swamped out by other effects of blood groups on fertility and mortality.

The tenfold difference between males and females in the rate of selection against type O illustrates how social forces affect natural selection. The difference arises because men reach their peak reproductive period later than women (a consequence of socially determined marriage ages), because men occupy more stressful occupations and develop duodenal ulcers more frequently, and because men die at a slightly higher rate than women at all ages. If the society were one in which the age at marriage is the same for the two sexes and the ages and causes of mortality are radically shifted, this difference would disappear. On the other hand, in a society in which infectious disease is the major cause of death, the selection against various blood types might be quite different. Bacterial pathogens have antigens on their cell walls that have some similarity to the ABO blood-group antigens. Thus, a type A person, lacking the anti-A antibody, may be more susceptible to infection by a bacterium whose cell walls carry an antigen that resembles the antigen of blood type A. There is, at present, no irrefutable evidence that people of different ABO blood types have differential disease resistance, but it is a reasonable possibility.

How Selection Works

The actual course of change in a population under the influence of natural selection is intimately related to the nature and amount of genetic variation. In the first place, no one can predict what changes will occur by selection simply by examining the relations between organisms and their environments. Selection is not "for body size," "for intelligence," or "for disease resistance." Selection is for fitness, for a greater reproductive rate. Whether that greater reproductive rate is realized in a population by an increase in body size, by an increase in disease resistance, by both, or by neither depends upon whether genetic differences in these characteristics exist among individual members of the population, as well as whether those genetic differences are associated with differential fitness. No selection for body size can occur in a population in which there is no genetic variation in body size. A major error in many attempts to reconstruct or predict human evolution is to postulate selection for characteristics without any evidence for (and sometimes with presumptive evidence against) the existence of genetic variation in those characteristics. Suggestions that such social traits as religiosity, dislike of strangers, and jealousy of one's own property are the outcome of direct selective pressures require that specific genetic differences for these traits once existed among human beings. Yet there is no evidence for any present genetic variation for atheism or any likelihood that genes could really act in this way.

In the second place, as a consequence of the absolute dependence of selective change on the existence of genetic variation, two different populations may experience selection for two very different traits even if the external forces of the environment are the same. Adaptation to cold stress can be accomplished either by producing extra heat, as in the Eskimos, or by becoming semidormant, as in Australian aborigines. Whether the calorie-producing or the calorie-conserving mode has evolved will have depended on which genes were varying in the ancestral population. There is more than one way to skin a cat or to warm a human.

Third, the same selective forces may result in different final expressions of the same characteristic, depending upon the nature of the genetic variation in the parental population. Some rhinoceros species have one short, stout horn, and some have two long, slender ones. We can reasonably suppose that horns evolved as a protection against preditors and competitors, but we are not obliged to explain why it is better to have one horn in India and two in Africa.

The African (left) and Indian (right) rhinoceros.

The time pattern of increasing frequency of a new favorable allele A that has entered a population of aa homozygotes.

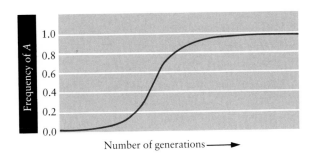

These are alternative outcomes of the same selective forces operating on two different ancestral populations with different genetic variation for growth patterns of hair (from which rhinoceros "horns" are made).

Fourth, the way in which selection works depends upon the quantity of variation as well as its quality. When nearly everyone in a population is genetically the same, selection is very ineffective in changing gene frequencies because there is not much variation to select from. As selection proceeds, the rare but favorable form becomes more and more common, with the result that there is more total genetic variation in the population. Selection than speeds up because of the greater variation. As more time passes, however, the favorable form becomes yet more common until finally nearly everyone is of that type. Again, variation has been lost, and selection slows down. The figure above shows that the S-shaped curve of selective change as a favorable character is at first rare, then of middling frequency, and finally overwhelmingly preponderant in the population. The principle that rate of change under selection is proportional to the amount of genetic variance has important consequences for the alternative outcomes of selection. If two different characteristics are genetically variable but one has much more genetic variance than the other, the one with more variance will evolve much more rapidly and selection will be "for" that character to the virtual exclusion of the less variable trait. A gene whose frequency in a population is 50% will increase in frequency 25 times as fast as one whose frequency is 0.1%, even if it is only one-tenth as favorable under natural selection. Occasional rare very favorable mutations are thus a less effective substrate for natural selection than more widespread variation of a less dramatic sort.

Fifth, as we are told in Ecclesiastes, "the race is not to the swift, nor the battle to the strong, neither yet bread to the wise . . . but time and chance happeneth to them all." Even to genes. The result of random genetic drift is that the outcome of the evolutionary process is uncertain, and sometimes it may be opposite to the general direction of selective forces. Mildly deleterious genes can and do become homozygous in populations by drift, despite selection. Mildly advantageous mutations, each time they appear, have a very low probability of spreading through the population. Even if the drunkard has a dim idea that he wants to walk south, he may easily wind up at the north end of the block because his staggering is so much stronger than his feeble directional movements; and if he begins right on the northern curb, he will almost certainly fall into the northern gutter immediately.

The Course of Evolution

The Yanomamo Indians of the upper Orinoco drainage in Brazil live in more than 100 small villages of roughly 100 persons each. New villages are formed by small emigrant groups, and there is some migration of people between villages. The Yanomamo as a whole are in contact with other tribes and, now, with an organized Western state. They derive originally, as do other American Indians, from the Asian migrants who crossed the Bering Strait land bridge and who reached South America about 10,000 years ago. The Yanomamo villages differ very little from each other in the frequencies of various genes, nor are they very different from such Amazonian tribes as the Xavantes. Like most South American Indians, they differ from Europeans in having a very low frequency of the ABO blood-group allele I^B and a high frequency of the Diego blood-group allele Di^a, which is totally absent in people of non-Asiatic origin. The process that formed these genetic similarities and differences in the past, and is still operating to mold them, is the same one that has operated over the whole history of our species and every other species of organism.

Within each local village population, sexual recombination and occasional mutations are constantly generating genetic variation. Because the populations are so small, however, genetic drift is powerful. Haplotypes of the HLA system, for example, are being lost in each village because, by chance, they have not been included in any effective gamete. Alleles of polymorphic genes, such as the Rh blood-group alleles, fluctuate in frequency from generation to generation within each village. Random drift tends to cause the populations to differentiate one from another because, for example, one HLA haplotype will become predominant in one village whereas a different haplotype will spread through the population of another. If the villages were totally isolated from each other, each would eventually become totally homozygous, but there would be large differences between villages. This centrifugal force of random drift causing the populations of villages to diverge from each other and to become internally homogenous is counteracted by migrations that bring lost alleles and haplotypes back into a village from neighboring groups and thereby make the different villages more like each other. Each force has an opposite effect on the variation within and between villages. Drift, which destroys variation within villages, causes

A Yanomamo Indian of South America. The Yanomamo, who have only recently been in contact with European society, have provided anthropologists and students of human evolution with a vast fond of information about human evolutionary differentiation.

differentiation between them. Migration and mutation, which add variation within populations, tend to iron out the differences between them.

These same forces are operating at higher levels of population composition. As a group, the Yanomamo are diverging from the Xavante because of genetic drift operating within each tribe. (This divergence is much slower than that between the people of two villages, because the Yanomamo tribe is altogether 10,000 people.) Without some migration from the outside, the Yanomamo would eventually become genetically homogeneous and different from their neighbors. On an even longer time scale, all South American Indians have diverged from their Asian ancestors, partly because of genetic drift. Presumably, the original emigrants from Asia who crossed the Bering Strait land bridge were fairly small in number and so, as they began their life in the Western Hemisphere, were genetically somewhat different from their ancestors as a group. This is probably when they lost the I^B allele, since it is so low in frequency among all American Indians. By chance, the frequency of the Diego blood-group allele Di^a may have been unusually high among these founding families.

At the same time that drift, migration, and mutation have been destroying and reconstituting variation, natural selection has been operating in different directions. Some selective forces, perhaps affecting body conformation and skin color, have been moving in the same direction in all tropical forest-dwelling Indians. This selection would have made the Yanomamo and Xavante peoples alike, but it differentiated them from the Plains Indians of the north from whom they were separated less than 500 generations ago. Simultaneously, there may be very specific selective forces that cause divergence between Yanomamo and and Xavante (or even between the people of two Xavante villages). Small dietary differences between tribes and differences in the mineral content of the water and food plants from one village to the next could be the basis of such differentiating selection. Moreover, the same selective forces operating in different villages may, nevertheless, cause the villages to diverge because, by chance, they began with different assortments of alleles. It must be remembered that new villages are formed by very small founding populations that, because of sampling, may be quite different initially from their neighbors. Selection can accentuate these differences because, as we have seen, the same selective force may have several alternative outcomes.

The effect of selection on genetic diversity is complex. If there is a rare variant of a gene whose frequency is increased by selection, the amount of variation begins to increase as the rare gene becomes more common. Eventually the amount of variation will again decrease and disappear as the new form replaces the old. Some kinds of natural selection actually preserve variation rather than using it up. If a gene has two alleles and *both* homozygotes are selected against

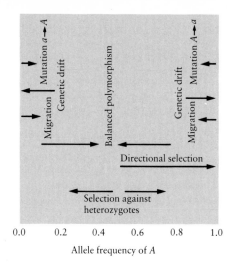

The effects of various evolutionary forces on the frequency of gene alleles.

relative to the heterozygote, then neither allele will replace the other, and a permanent stable polymorphism will be produced. This form of balancing selection is widely talked about but rarely seen. The classic—and only known—case in humans is sickle-cell anemia: Most homozygotes for hemoglobin S die from anemia; some homozygotes for hemoglobin A die from malaria; but the heterozygotes survive both risks. If such balancing selection is common, it may explain much of the polymorphism that is observed in humans, but examples have not been easy to find. In general, the cause of most human polymorphism remains a mystery. It may simply be the balance between the forces of drift destroying variation and the forces of mutation and migration restoring it. The figure at the left summarizes how the different forces of mutation, migration, selection, and drift balance each other in molding genetic variation.

The Unity of the Human Species

Above all, the Yanomamo are human beings who share with all other human beings the historical outcome of selective and random forces that operated long in the past. The human species is young, perhaps not more than 10,000 generations old, and the major geographical races diverged from each other about 1500 generations ago, at most. The processes that cause gene-frequency change are slow. Once lost, an allele may not easily be replaced again in 1500 generations in a population of, say, 1000 people. Selective forces, even if they are ten times as strong as those that we calculated for the ABO blood group and duodenal ulcers, will have changed gene frequencies very little since the origin of our species. On the other hand, a very small amount of migration—as little as one migrant individual exchanged between groups in each generation—is quite sufficient to prevent differentiation between groups by genetic drift. The bulk of the genetic polymorphisms among the Yanomamo and nearly all of their constant genes are widely shared with other human groups, as we saw in Chapter 8. The unifying forces of migration and common selection have kept human beings all over the world as members of the same species, despite the differentiation that occurred when humans were widely dispersed, living in very small, isolated populations. This need not have been. Had we been less mobile and less adaptable, both as individuals and as cultures, the disruptive processes of local natural selection and drift might have fragmented our species into local units that would become more and more divergent from each other and, in time, might even have formed different species. If anything is clear about the direction of human evolution, it is that the early differentiation of people into local groups, while still very much a part of our biological diversity, is on the decline. The unifying forces of migration and of common selection through common environment and common culture are stronger than they have ever been.

The Human Past

Reconstructing the evolutionary past of the human species is almost as difficult as predicting the future, although both are common exercises that biologists engage in, especially when they address a nonscientific public. All claims that human societies used to be like this or like that need to be looked at with the greatest skepticism. Virtually all such claims are pure speculation.

When we consider the remote past, before the origin of the actual species *Homo sapiens,* we are faced with a fragmentary and disconnected fossil record. Despite the excited and optimistic claims that have been made by some paleontologists, no fossil hominid species can be established as our direct ancestor. For many years, it was thought that Neanderthal man was an early form of *Homo sapiens,* but it is now probable that the Neanderthal was a separate species living at the same time as *Homo sapiens,* as recently as 30,000 years ago. The earliest forms that are recognized as being hominid are the famous fossils, associated with primitive stone tools, that were found by Mary and Louis Leakey in the Olduvai gorge and elsewhere in Africa. These fossil hominids lived more than 1.5 million years ago and had brains half the size of ours. They were certainly not members of our own species, and we have no idea whether they were even in our direct ancestral line or only in a parallel line of descent resembling

Three fossil skulls from East Africa, thought to be *Australopithecus robustus* (upper), *Australopithecus africanus* (middle), and an early form of *Homo* (lower).

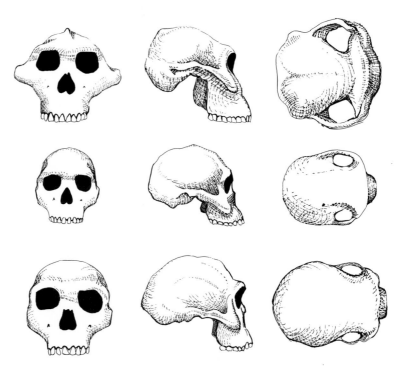

our direct ancestor. Assertions that we are descended either from a large, vegetarian, apelike ancestor *(Australopithecus robustus)* or from a smaller, carnivorous one *(Australopithecus africanus)* and that we owe our present natures to the eating habits of these early "ancestors" are totally without merit. We have not the faintest idea of which of these species—if either—is in the direct line of human descent. All the attempts to assert that one or another fossil species is our direct progenitor reflect an outdated notion that evolution is strictly linear and that all fossil forms must be fitted somewhere along a single sequence connecting the past with the present. In fact, evolution occurs by a process of repeated branching, with most of the branches becoming extinct fairly rapidly. If new species arise fairly frequently and most last for only a relatively short time before becoming extinct, there will be, at any moment, a large number of parallel evolutionary lines stemming from a common ancestor, only one of which may be represented in the distant future (see the figure at the left), all of the rest having become extinct. A major problem in reconstructing human evolution is that we have no close living relatives. The chimpanzee and the gorilla were connected to us by a common ancestor at least 7 million years ago, so that more than 14 million years of independent evolution must be traversed in tracing up from these apes to that common ancestor and then back down to us. To keep that amount of time in evolutionary perspective, we should note that the mammals as a whole are only 140 million years old.

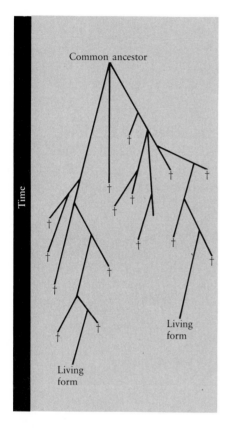

A typical highly branched pattern of evolution showing that at an earlier time many related species existed simultaneously but most have become extinct.

Gorilla and chimpanzee, our nearest (but not very near) living relatives.

In passing from the evolution of human ancestors to the history of our species itself, in passing from paleontology to archaeology, we pass from a concentration on morphological evolution to an interest in cultural change. In particular, we want to know how human cultural organization—including population size, family structure, migration patterns, production methods, habits of life, and causes of disease—may have helped to determine, and been determined by, biological evolution. What has natural selection been doing to the human species for the past 30,000 years? The problems of reconstruction here are just as great as they are for the fossil record. Archaeology depends upon finding the "hard parts" of human culture just as paleontology depends upon finding bones. But hard parts—stone, pottery, metal, even wood and clothing—are left in profusion only by highly organized, sedentary, densely populous civilizations, all of which are fairly recent. Most archaeological finds have been of agriculture settlements, towns, cities, and organized city states, which began about 8000 B.C., when cereals were first cultivated in the Near East. That is, most archaeological reconstructions are of civilizations very much like ours. The preagricultural hunters and gatherers who characterized the human species for all of its history prior to about 10,000 years ago, and whose time on earth encompassed most of human evolution, have left us precious little to judge them by. All that are left are some stone and bone implements, some fireplaces, some kitchen middens, and some extraordinarily beautiful wall paintings of animals

Preserved hard parts of past cultures: Iron Age tongs and Pleistocene paleolithic and recent neolithic implements.

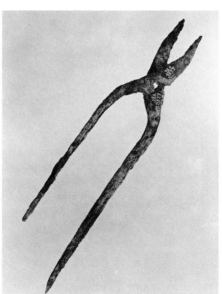

A Kalahari Bushman.

and hunters. None of those artifacts are very plentiful, because their makers were not very plentiful. None of them allow us to reconstruct age distributions, marriage patterns, main causes of morbidity and mortality, fertility rates, divisions of labor within families, or any social structures. About all that can be said is that there weren't very many people but the species as a whole (although not necessarily any individual person or family) could move very rapidly. It took less than a thousand years for the early migrants from Asia to move from what is now western Canada to South America, a distance of 5000 miles.

The alternative to digging up the human past has been to reconstruct it by analogy with the lives of present-day hunting and gathering people. The Ituri Forest Pygmies, the Kalahari Bushmen, the Australian aborigines, the Eskimos, and the North American Indians are regarded as relics of the human past whose present-day cultures can be used as models of past social relations. This is a dangerous method. Few (some anthropologists say none) of the present-day "primitive" people have not been influenced by state-organized societies. One should recall that the horse, which played such a central part in the lives of the Plains Indians, was introduced into North America by the Spanish Conquistadores only 350 years before the battle of Little Big Horn. Moreover, at least some modern-day hunters and gatherers are probably recent refugees from

De Soto's discovery of the Mississippi.

more hospitable environments from which they were driven by more aggressive and successful people, and so their culture is a relatively new adaptation to their circumstances. The view that the Bushmen are "primitive" is like the notion that bacteria are primitive. One may forget that the bacteria have undergone a longer evolutionary history than the vertebrates. Certainly the evolutionary ancestors of all organisms must have been single-celled organisms, but precisely which features of a modern bacterial cell are vestiges of a billion years of evolution and which have evolved recently is not always clear. Thus, we cannot know from studying the Bushmen what the average life expectancy may have been during the Neolithic. Most modern-day hunters and gatherers are living marginal existences precisely because the more agreeable parts of the world have been taken over by agricultural-technical cultures.

Ironically, the best evidence we have about overall human migration patterns in the past comes from the present distribution of gene frequencies; so it is the genetic facts that are used to infer the evolutionary processes, rather than vice versa. The maps on the next page show the present clinal pattern of blood-group frequencies in the Old World of blood types O and B. We see lines, radiating out from central Asia, that no doubt represent the progressive spread of people eastward and westward. This same phenomenon, on a smaller scale, is shown for the Japanese islands on page 116. It is the weakness of natural selection and the recency of the spread of the human species that allows us to use these distribution maps as pictures of migration.

The reconstruction of the forces of natural selection on the human species is entirely a matter of speculation, because we know virtually nothing about the heritability of most human characters, or about the differential reproductive rates associated with them, or about the way in which that differential may have changed with time. An illustration of these problems is provided by the example of selection for human birth weight, one of the best documented cases of differential survival for a continuously varying trait. The figure at the left shows the birth-weight distribution of English children and the curve of infant survival for the different birth-weight classes. It is clear that selection favors infants of intermediate weight and that there is a close correspondence between the optimum weight and the mean weight. For natural selection to be actually having an effect on the evolution of the species, however, there must be heritability of birth weight. An estimate of heritability of birth weight, made from comparisons of monozygotic and dizygotic twins, is .63. Because monozygotic twins often share the same placenta, whereas dizygotic twins do not, it is impossible to say whether the estimate of heritability is really only a measure of environmental similarity. Still, it is likely that birth weight has some heritability, especially within a population that does not have wide variation in maternal nutrition.

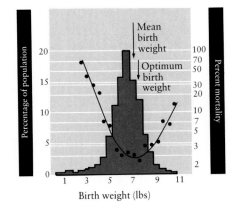

The relation between birth weight and survival (curve) superimposed on the actual distribution of birth weights of English infants.

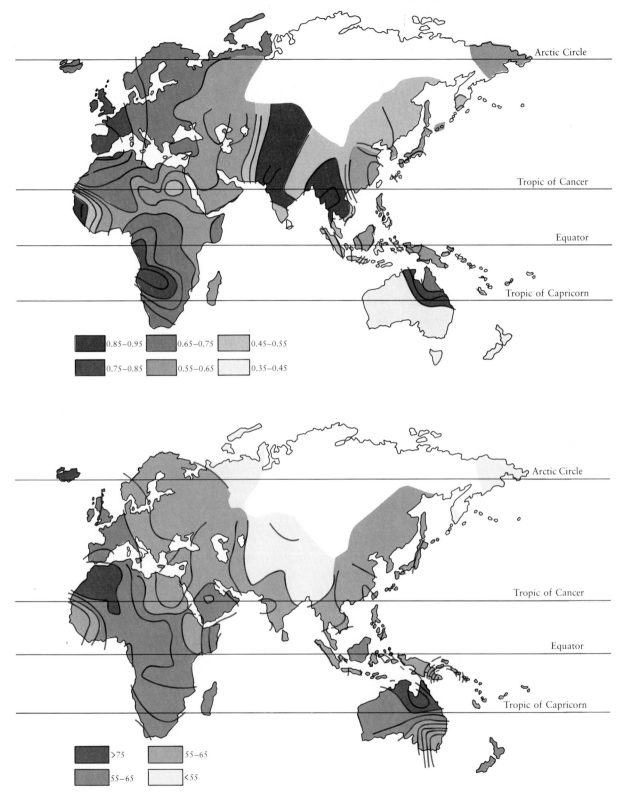

Arctic Circle

Tropic of Cancer

Equator

Tropic of Capricorn

| | 0.85–0.95 | | 0.65–0.75 | | 0.45–0.55 |
| | 0.75–0.85 | | 0.55–0.65 | | 0.35–0.45 |

Arctic Circle

Tropic of Cancer

Equator

Tropic of Capricorn

| | >75 | | 55–65 |
| | 55–65 | | <55 |

Maps showing the present-day clines in the frequencies of blood type O (top) and A (bottom) in the Old World. These clines radiate out from Central Asia and probably represent a trace of the outward migration of early humans to occupy the rest of the land mass.

What cannot be judged easily, however, is whether the same selective force operated in preindustrial society, when there were different patterns of maternal activity, different obstetrical practices, different nutritional patterns, and different adult body sizes. Thus, we do not really know how birth weight was selected in the past.

It is reasonable to guess that selection has always operated in favor of resistance to infectious disease. The various antigenic polymorphisms that are so common in all human populations may very well be the result of that selection, inasmuch as it seems likely that different blood-group antibodies may react with different antigens on bacterial cell walls. As we have seen, it is probable that selection has also operated to modify body conformation as a means of effecting temperature control. A complication of that process has been selection for the efficient storage of fat, both as a source of metabolic water and as a way to store energy to ensure survival when food supplies are unreliable.

Aside from such physiological traits, it is easy to make a list of behavioral characteristics that are advantageous to human life and for which one can construct imaginative scenarios of selection. Thus, the willingness to cooperate in food gathering and distribution, a striking characteristic of modern hunting and gathering groups, is clearly of advantage to individual people and their families. Noncooperative people would be excluded from the group and probably would starve, because both hunting success and the leveling out of the vagaries of individual bad luck require group solidarity. The problem with making up such stories about selection is that there is no limit to them, and there is not the slightest evidence that there is (or ever was) any genetic variation that affects cooperativeness. Are there genotypes that cause some people to be less cooperative than others? Was there a time when humans were not cooperative because they lacked the appropriate genes, which then arose by mutation and recombination and were selected for? It seems much more likely that cooperation is an adaptive cultural response by an extremely clever species to the perceived uncertainty of the environment. If anything has happened to natural selection in the past, it is that it has become less and less important as genetic differences between people have become less and less relevant to their individual survival.

The Human Future

The only certainty about the future of our species is that it is limited. Of all the species that have ever existed, 99.999% are extinct. The average lifetime of a carnivore genus is only 10 million years, and the average lifetime of a species is much shorter. Indeed, life on earth is nearly half over: Fossil evidence shows that life began about 3 billion years ago, and the sun is due to become a red giant about 4 billion years from now, consuming life (and eventually the whole earth) in its fire.

It is remarkably difficult to make any reliable predictions about our future between now and the uncertain date of our extinction, because human evolution depends so much on the state of cultural organization. It is very likely that we will continue to be a single species, because the forces of genetic cohesion, especially migration and uniform selection, seem to be increasing. It is hard to imagine a universal catastrophe of such a nature that the species would be broken up into isolated small groups for the tens of thousands of years that would be required for the formation of separate species. We would need not only to be returned to the Stone Age but to be deprived of all the knowledge about the physical world that has been accumulated since then. Total extinction seems more likely.

The force of natural selection, in general, must become even weaker than it now is, as social institutions iron out the effects of individual biological variation. The drop in the death rate from tuberculosis in Europe, from 4000 per million in 1840 to 13 per million now, clearly means that selection for resistance to tuberculosis has virtually ceased (if there ever was any genetic variation for it in the first place). In fact, the opportunity for selection of *any* kind is rapidly decreasing in technologically advanced countries for purely demographic reasons. For there to be selection, there must be some variation in family size. If every person born were eventually married and each couple were to produce exactly two children, there would be no scope for selection because there would be no differential reproductive rate among individual people. Although we have certainly not reached a stage of no variation in reproductive rates, there has been a strong trend in this direction. In 1900, the variance in family size was three to four times as great as it is at present in the industrial countries of the West. There are also great differences between countries in this respect. In 1950, Venezuela's variance in family size was 19, whereas it was only 5 in the United States. More to the point is the variation between regions in reproductive rate. Countries of the Southern Hemisphere are contributing many more genes to the species as a whole than are northern industrial countries. The consequence will be that the human species will come, more and more, to have gene frequencies that at present characterize Africans, South Americans, and South Asians. As far as anyone knows, this trend will have no interesting consequence, except that it will make the species as a whole look less variable but darker in skin color.

The fear that the human species will be biologically inundated by "lesser breeds without the law" has been a preoccupation of intellectuals, especially in Anglo-Saxon countries. The founders of modern statistics, Francis Galton, Karl Pearson, and R. A. Fisher, were all convinced that the intelligence of the species was decreasing because the people of the lower classes who were presumed to

Average number of offspring per person in relation to IQ

IQ Range	Number Reporting	Number of Offspring per Person
≥ 120	82	2.598
105–119	282	2.238
95–104	318	2.019
80–94	267	2.464
69–79	30	1.500
Total sample	979	2.236

Source: C. Bajema, *Eugenics Quarterly* 10(1963): 175–187.

have lower IQs, were outbreeding their betters. Statistics showing higher reproductive rates for people with lower IQs were widely accepted until they were shown in 1963 to be a statistical artifact resulting from counting only families with at least one child. All the childless people had been left out. When the complete data were taken, the results were quite different, as shown in the table at the left. From these results, one cannot predict any evolutionary trend in IQ, even if IQ were heritable, which is by no means certain.

Like the reconstruction of past selection, the prediction of future selection is limited only by one's imagination and one's willingness to make statements unsupported by evidence. No one who seriously contemplates the extraordinary changes that take place in every aspect of human life during brief periods of history can pretend to guess the biological future. Only 100 human generations separate us from the foundation of the Roman Republic. In the first 200 years after the Hegira, the people of Arabia and North Africa went from a backward pastoral existence to the heights of culture and world power, dominating the Mediterranean world in art, science, poetry, mathematics, and political skill. When Paris was a village on an island in the Seine, Cordoba under the Umayyids was the center of Western civilization. Given the great genetic similarity among human groups and the very slow rate of genetic change produced by selection, as contrasted with the remarkable diversity of human cultures and the almost instantaneous changes that occur in history, it is difficult to see what the relevance of human biological diversity is to the future of human life. Biologists often point to sickle-cell anemia as an example of how natural selection operates in human populations. Yet nothing better illustrates the hegemony of culture. The process of selection that has kept the allele for hemoglobin S common in parts of Africa was completely reversed for large numbers of Africans by a political and economic event: the enslavement of those Africans and their transport to America. A second political event, the decision of the World Health Organization to eradicate malaria, began to change the nature of selection in Africa too, until a third political event, a wholesale retrenchment of the antimalaria campaign, restored the biological status quo. In America, where there is no advantage to being heterozygous for the hemoglobin S allele and where natural selection is operating to eliminate the allele through the death of children with sickle-cell anemia, a fourth political event, the demand of blacks for social power, has resulted in a modest research effort to find a treatment for the disease.

It is common, in books about human biology or evolution, to end them with a grand, and sometimes pompous, disquisition on the future of the human species and the meaning of its existence. Beginning with the truism that human beings are animals and have evolved from yet other animals, biologists often end with the false assertion that human beings are merely animals and can be understood by extrapolation from monkeys, wolves, and grey-lag geese. The chief purpose of this book has been to make clear that such a view is wrong. The most striking property of human biology—a property that is, indeed, the outcome of human biological evolution—is that human beings make their own individual and collective futures. The evolution of the human central nervous system, and the evolution of the hand, eye, and tongue that it entails, has freed human beings to an extraordinary degree from biological constraints that are common to our animal relatives and ancestors. Human consciousness and human social organization are the organs with which we determine our individual and collective natures, both present and future. The great evolutionist Theodosius Dobzhansky wrote that "nothing in biology makes sense except in the light of evolution." But we must add that "nothing in human evolution makes sense except in the light of history."

Sources of Illustrations

page xii
Photography by Jan Lukas, © 1980/Photo Researchers, Inc.

page 2 and page 3 (left)
Photography by Guy Gillette/Photo Researchers, Inc.

page 3 (right)
Photography by Henri Cartier-Bresson/Magnum.

page 5 (left)
Photography by David Hurn/Magnum.

page 5 (middle)
Photography by Philippe Halsmann/Magnum.

page 5 (right)
Courtesy of the Tamiment Collection. From *A Pictorial History of American Labor* by William Cahn. Copyright © 1972 by William Cahn. Used by permission of Crown Publishers, Inc.

page 8
Adapted from a graph in *Science for People,* Winter, 1977–78.

page 10 (left)
Photography by Henri Cartier-Bresson/Magnum.

page 10 (right)
Photography by Rene Burri/Magnum.

page 11 (top)
Photography by O. Monsen/courtesy of the United Nations.

page 11 (bottom)
Photography by Marc Riboud/Magnum.

page 12 (left)
Photography by John Laundis/Black Star.

page 12 (right) and page 13 (left)
Courtesy of the United Nations.

page 13 (right)
Photography by Nicola Giordano/Black Star.

page 15
From *Land of Lilliputians.* [Jonathan Swift], *Travels into Several Remote Nations of the World,* vol. 1, by Lemuel Gulliver [pseud.]. London: B. Motte, 1727. Rare Books and Manuscripts Division, The New York Public Library; Astor, Lenox, and Tilden Foundations.

page 17 (top)
Courtesy of E. B. Lewis, California Institute of Technology.

page 17 (bottom) and page 21 (left)
Adapted from drawings in the second edition of *An Introduction to Genetic Analysis* by David T. Suzuki, Anthony J. F. Griffiths, and Richard C. Lewontin. W. H. Freeman and Company. Copyright © 1981.

page 21 (middle and right)
Courtesy of Anthony J. F. Griffiths.

page 23
Adapted from a drawing in *Experimental Studies in the Nature of Species,* vol. 3, by Jens Clausen, David D. Keck, and William M. Hiesey, Carnegie Institution of Washington.

page 24
Photography by Ira Berger.

pages 27 and 28
Adapted from drawings in the second edition of *An Introduction to Genetic Analysis* by David T. Suzuki, Anthony J. F. Griffiths, and Richard C. Lewontin. W. H. Freeman and Company. Copyright © 1981.

page 29
Courtesy of Patricia Farnsworth.

page 38 (right)
Electrophoretic pattern of hemoglobin courtesy of Anthony C. Allison.

page 43
Micrograph by Dr. Shettles/courtesy of The American Museum of Natural History Photo Researchers, Inc.

page 44
After photographs published in the *Journal of Heredity* in an article by C. B. Davenport.

page 46 and page 47 (top)
Electron micrographs by C. J. Marchant and A. M. Adamovich.

page 50
Space-filling model is from "Gene Structure and Protein Structure" by Charles Yanofsky. Copyright © 1967 by Scientific American, Inc. All rights reserved. Unwound section is adapted from an illustration in "The Synthesis of DNA" by Arthur Kornberg. Copyright © 1968 by Scientific American, Inc. All rights reserved.

page 51
Adapted from an illustration in "The Structure of the Hereditary Material" by F. H. C. Crick. Copyright © 1954 by Scientific American, Inc. All rights reserved.

page 53
Adapted from an illustration in "The Genetic Code: III" by F. H. C. Crick. Copyright © 1966 by Scientific American, Inc. All rights reserved.

page 55
Electron micrograph by O. L. Miller, Jr., and Barbara A. Hamkalo.

pages 58 and 61
Adapted from drawings in the second edition of *An Introduction to Genetic Analysis* by David T. Suzuki, Anthony J. F. Griffiths, and Richard C. Lewontin. W. H. Freeman and Company. Copyright © 1981.

page 62
Data on Pygmies and Dinkas are from *Genetics and the Races of Man* by William C. Boyd, D. C. Heath, 1950.

page 63 (margin)
Photography by George Rodger/Magnum.

page 63 (top left)
Photography by George Holton/Photo Researchers, Inc.

page 63 (top right)
Photography by Peter Vandermark/Stock, Boston.

page 66 and page 69 (top)
Adapted from graphs in *Principles of Human Biochemical Genetics,* 2d rev. ed., Harry Harris, editor, North-Holland, 1975.

page 74
Photography by Kathryn Abbe, © 1979.

page 77
Music Division, The New York Public Library at Lincoln Center; Astor, Lenox, and Tilden Foundations.

page 78 and page 79 (margin)
From the collection of Kathryn Abbe and Frances McLaughlin-Gill.

page 79 (top)
Photography by Kathryn Abbe, © 1979.

page 80
Photography by Ira Berger.

page 81
From *Phrenological Illustrations* by George Cruikshank. Courtesy of The New York Academy of Medicine Library.

page 84
Post-Dispatch Pictures/Black Star.

page 85 (left)
Photography by Henri Cartier-Bresson/Magnum.

page 85 (right)
Courtesy of The American Museum of Natural History.

page 86 (left)
Photography by B. Bhansali/Black Star.

page 86 (right)
Courtesy of the United Nations.

page 87 (left)
William Williams Papers, Rare Books and Manuscripts Division, The New York Public Library; Astor, Lenox, and Tilden Foundations.

page 87 (right)
Photography by Casper Mayer/courtesy of The American Museum of Natural History.

page 89 (left)
Photography by Mary Lloyd Estrin.

page 89 (right)
Photography by Lisa Barlow.

page 90
Illustration by George Cruikshank from *The Adventures of Oliver Twist: or The Parish's Boy's Progress* by Charles Dickens. London: Bradbury and Evans, 1846. Rare Books and Manuscripts Division, The New York Public Library; Astor, Lenox, and Tilden Foundations.

page 91
From *General Psychology* by Henry E. Garrett, © 1955. Reprinted by permission of D. C. Heath and Company.

page 92
Adapted from Figure 35 on page 216 of *Essentials of Psychological Testing,* 2d ed., by Lee J. Cronbach. Copyright 1949 by Harper & Row, Publishers, Inc. Copyright © 1960 by Lee J. Cronbach. Courtesy of the publisher.

page 96
Adapted from graphs published in an article by S. Bowles and V. Nelson in *The Review of Economics and Statistics,* February, 1974.

page 99
Adapted from an illustration in "Genetics and Intelligence: A Review" by L. Erlenmeyer-Kimling and L. F. Jarvik, *Science* 142(1963):1477–79. Copyright 1963 by the American Association for the Advancement of Science.

page 104
Lewis W. Hine Collection; U.S. History, Local History, and Genealogy Division, The New York Public Library; Astor, Lenox, and Tilden Foundations.

page 105
Photography by J. Pavlovsky/Sygma.

page 106 (top)
Engraving by T. Phillibrown/The Granger Collection.

page 106 (bottom)
Engraving from a sketch by W. W. Bode/The Granger Collection.

page 107 (left)
The Rape of the Sabine Women by Nicholas Poussin (1594–1665). The Metropolitan Museum of Art, Harris Brisbane Dick Fund, 1946.

page 107 (right)
Sinclair Hamilton Collection of American Illustrated Books, Princeton University Library.

page 109
Photography by Scheler/Black Star.

page 110
Adapted from a graph in *Genetics, Evolution, and Man* by W. F. Bodmer and L. L. Cavalli-Sforza. W. H. Freeman and Company. Copyright © 1976.

page 111
Photography by Gary Azon, © 1982.

page 112
Adapted from maps in the *Times Atlas of World History.* London: Times Books Limited.

page 115 (upper left)
Photography by John Messina/Black Star.

page 115 (upper right)
Photography by Bill Grimes, © 1979/Black Star.

page 115 (lower left)
Photography by Pat Goudvis, © 1978/Black Star.

page 115 (lower right)
Photography by Joe Covello/Black Star.

page 116 (top) and page 118
Adapted from maps in *Genetics and the Races of Man* by William C. Boyd, D. C. Heath, 1950.

page 116 (bottom)
Adapted from a map in *The Distribution of the Blood Groups in the United Kingdom* by A. Kopec, Oxford University Press, 1970.

page 124 (left)
Photography by Henri Cartier-Bresson/Magnum.

page 124 (right)
Photography by Ingeborg Lippman/Magnum.

page 125 (top)
Adapted from a map in *Razze e popoli della terra* by R. Biasutti, Turin: UTET, 1951.

page 125 (lower left)
Photography by René Burri/Magnum.

page 125 (lower middle)
Photography by George Rodger/Magnum.

page 125 (lower right)
Photography by Hubertus Kanus, © 1974; Rapho/Photo Researchers, Inc.

page 126 (upper left)
Photography by John Running, © 1978/Black Star.

page 126 (upper right)
Courtesy of The American Museum of Natural History.

page 126 (lower left)
Photography by Philip Jones Griffiths/Magnum.

page 127 (upper left)
Photography by Peter Thomas/Black Star.

page 127 (upper right)
Photography by M. Rothstein/courtesy of the United Nations.

page 127 (lower left)
Adapted from a graph in "The Hemoglobinopathies Thalassemia" by H. Lehmann, R. G. Huntsman, and J. R. Ager in *The Metabolic Basis of Inherited Disease,* 2d ed., by J. B. Stanbury, J. B. Wyngaarden, and D. S. Fredrickson, editors. New York: McGraw-Hill. Copyright © 1966.

page 129
From "The Distribution of Man" by William W. Howells. Copyright © 1960 by Scientific American, Inc. All rights reserved.

page 134
Photography by Harry Wilks, © 1981/courtesy of the Mannes School of Music.

page 135 (left)
The Bettmann Archive, Inc.

page 135 (middle)
Photography by Cathy Stanley/courtesy of the publishers of *Mountain Life & Work,* Clintwood, Virginia.

page 135 (right)
Lewis W. Hine Collection; U.S. History, Local History, and Genealogy Division, The New York Public Library; Astor, Lenox, and Tilden Foundations.

page 136 (top)
Courtesy of Fred J. Dill.

page 141 (left)
Photography by Robert Smith/Black Star.

page 141 (right)
Photography by Bob Vose/Black Star.

page 144 (upper left)
Photography by René Burri/Magnum.

page 144 (upper right)
Courtesy of the United Nations.

page 144 (bottom) and page 145 (bottom)
Photography by Elliott Erwitt/Magnum.

page 145 (top)
Courtesy of the United Nations.

page 147
Thames and Hudson Ltd.

page 158 (left)
Photography by Charles Marden Fitch/Taurus Photos.

page 158 (right)
Photography by Irene Vandermolen; Leonard Lee Rue/Tom Stack & Associates.

page 159
Adapted from a graph in the second edition of *An Introduction to Genetic Analysis* by David T. Suzuki, Anthony J. F. Griffiths, and Richard C. Lewontin. W. H. Freeman and Company. Copyright © 1981.

page 161
Photography by Flavio Damm/Black Star.

page 162
Adapted from a drawing in the second edition of *An Introduction to Genetic Analysis* by David T. Suzuki, Anthony J. F. Griffiths, and Richard C. Lewontin. W. H. Freeman and Company. Copyright © 1981.

page 163
Adapted from an illustration in "The Hominids of East Turkama" by Alan Walker and Richard E. F. Leakey. Copyright © 1978 by Scientific American, Inc. All rights reserved.

page 164 (lower left and right)
Courtesy of the Philadelphia Zoo.

page 165 and page 166 (top)
Courtesy of The American Museum of Natural History.

page 166 (bottom)
The Granger Collection.

page 167
Adapted from a graph in *The Genetics of Human Populations* by L. L. Cavalli-Sforza and W. F. Bodmer. W. H. Freeman and Company. Copyright © 1978.

page 168
Courtesy of D. E. Schreiber, IBM Research Laboratory, San Jose, California.

Index